DATE DUE

HELP YOUR CHILD LEARN NUMBER SKILLS

Frances Mosley
Primary schools mathematics adviser
and
Susan Meredith

Designed by Mary Cartwright

Illustrated by Tony Kenyon and Teri Gower
Photographs by Lesley Howling and Bob Mazzer
Text edited by Robyn Gee

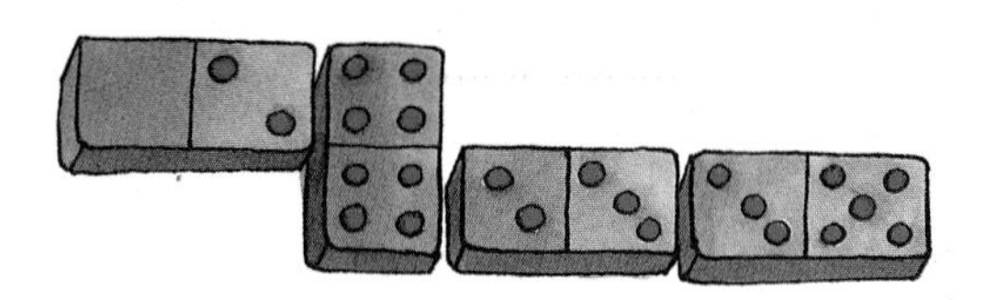

CONTENTS

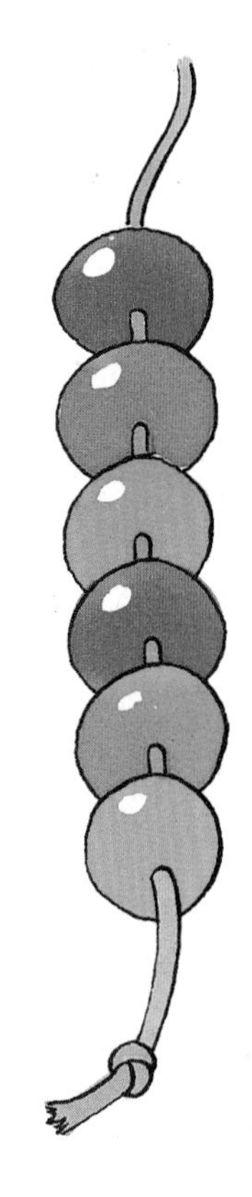

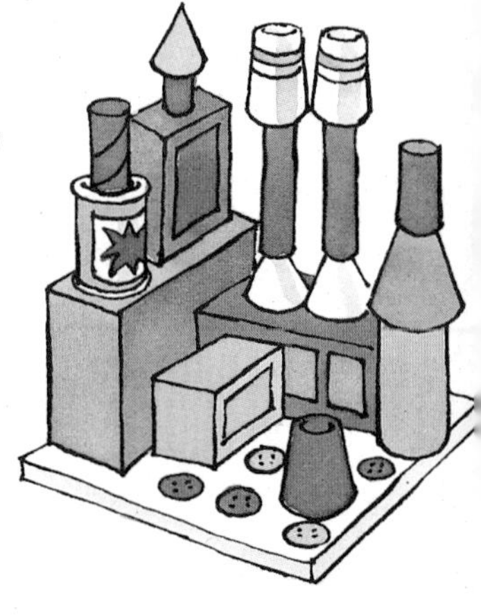

ABOUT THIS BOOK

This book aims to help parents encourage their children to enjoy math from the very beginning at home and during their early years at school. It is mainly about number skills but also deals with the other aspects of math, such as shape and measurement, which pre-school and primary school children come across.

Many parents question their ability to help their children with math because they were not very good at the subject themselves at school. This book is not about sitting down and doing sums with your child, however. It is aimed at helping children to develop a positive attitude to math and there is plenty you can do to encourage this in an informal way. Many basic mathematical ideas can be picked up naturally and at an early age in normal daily life.

Throughout the book the emphasis is on helping children to think for themselves and to find their own solutions to mathematical problems. Children learn better this way and remember more of what they have learnt. They also learn best if they are having fun, which is why there is so much emphasis on learning through play and games. There is more about how children learn math on pages 44-45. You may find it useful to read that section before you embark on some of the math activities with your child.

Use whichever of the ideas in the book appeal both to you and your child and don't feel you have to sit down for long math sessions together. Most of the activities are suitable for odd moments and your child will learn more from several short sessions than from just a few long ones. If your child shows no interest in a certain activity, don't force it upon her. She may not be ready for it yet or she may simply not like it. Above all, don't become over-concerned about your child's rate of progress. No precise guidelines have been given on the suitable age for each activity because the rate of children's mathematical development varies widely. Always remember that your child is most likely to learn if you are enjoying yourselves together.

STARTING TO COUNT

Young children soon become aware of numbers and often learn to recite them quite effortlessly. This is not the same as being able to count, but they will gradually learn to count too just so long as they see you counting in the course of everyday life and if you make a point of letting them help you count, for useful purposes in real situations. They will learn all the better if you are relaxed and casual about any counting you do together.

Number rhymes and songs

One, two, three, four, five,
Once I caught a fish alive...

A good way of helping young children to learn the sequence of numbers, both forwards and backwards, is through number rhymes and songs. Don't be at all concerned about mistakes and always remember to use the rhymes and songs because they are fun, not because you are deliberately trying to teach your child numbers. The words of some rhymes and songs are given on page 47. It is also worth trying out books and tapes of number rhymes and songs.

One-to-one correspondence

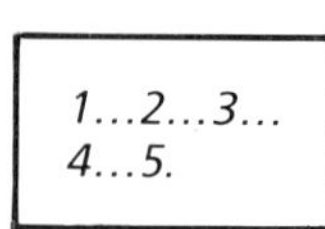

Even when young children can recite numbers in the right order, they often don't match number words to objects correctly when they try to count. This is because they have not grasped the notion of "one-to-one correspondence".

There are two aspects to this. The first is that one object goes with one other object, when you are pairing, for example cups with saucers. Later comes the idea that one number word goes with one object, when you are counting.

Pairing

To help your child with one-to-one correspondence, take advantage of any opportunities for her to pair objects.

One for Susie,
one for you,
one for me.
We've got too many.

When you are laying the table, for example, ask her to help you put out one cup for each person, one spoon for each bowl. Talk about whether there are enough of things, and then check. This will lead into the ideas of "too many" and "too few".

Toys which involve pairing, such as hammer boards and vehicles with peg people can give even very young children a practical introduction to one-to-one correspondence.

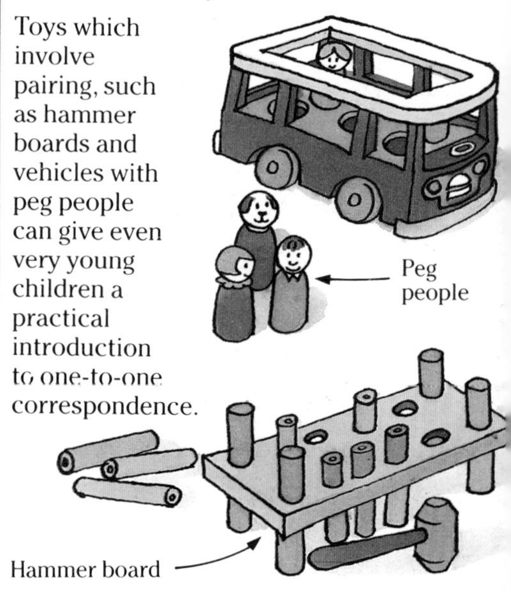

Counting for real

Now how many cups do we need? There's Laura, Ben and me. That's 1, 2, 3.

The best way to encourage your child to count is to let her see that you do it and that it is useful. When you need to count, do it aloud rather than in your head.

We'll need five apples. Can you help me count them? 1...

Involve your child by letting him count with you. Stick to small numbers of things at first and always point closely to, or touch, each one and perhaps move it aside as you count.

If he likes counting, he will find numerous opportunities for practice; cars parked at the curb, steps to climb, cans in the shopping basket.

4
Wait a minute. Is that right? Let's just check it.

Don't worry about mistakes either of matching numbers to objects or of number sequence. Don't feel you have to correct every error and when you do, try to be tactful.

Who has more?

Put out about six saucers, each with a different number of things in, for example, six used toothpicks, five pasta shells, four small buttons, three big buttons, two peas, one pebble.

Each choose a saucer and guess, without counting out, who has more. Check by pairing up the contents of the two saucers.

Variations

- Put the same number of things in two of the saucers.
- Put four big things in one saucer and four small things in another, say bottle tops and lentils.
- Put out bigger numbers of things.

Besides helping your child to understand "more", "less", "fewer" and the same, and that number is quite different from size, this game will also help her to start recognizing small numbers of objects instantly without counting out.

Grab more or fewer

First make a more/fewer spinner like the one shown on the right. Then give each of you a dish with 10 small objects in it.

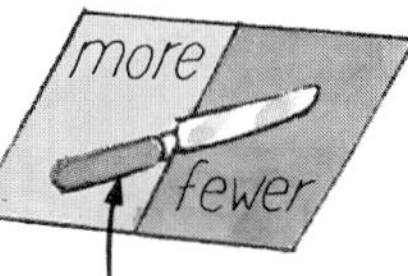

Each grab a handful from your dish and put them in front of you. Then spin the spinner. This tells you whether the person with more or fewer wins. Find out by pairing up the items.

Variation

- The loser of each game has to give one of his items to the winner. The one with the most in his dish when you stop playing is the overall winner.

Conservation of number

Young children do not realize that the number of things in a group does not change when they are re-arranged. If you count out a row of sweets with your child and then re-arrange them by putting them in a circle or by taking a sweet from one end of the row and placing it at the other end, your child may well need to count them again to be sure of how many there are. This is because he hasn't yet understood about "conservation" of number.

There is not a great deal you can do to teach the idea of conservation. It is one that children acquire gradually as they gain experience with counting and using numbers.

Conservation patterns

With your child, choose a low number, say four; then make as many patterns of four as you can, using say, bottle tops, sticky paper shapes or small pieces of fabric, and stick them down on card or paper. Make some of your patterns really spread out and others close together, some neat and others random.

This may help to get your child thinking about the "fourness of 4" as well as helping with instant recognition of groups of 4.

Reading numbers

You can help your child become aware of the shapes of numbers by talking about the numbers you see around you: on doors, buses, birthday cards, calendars, clocks, microwave ovens, televisions and telephones as well as in books. She will probably enjoy spotting numbers that have a special significance such as her age or the number of her flat or house. A clear, colourful number frieze on the bedroom wall will also help to familiarize her with numbers.

Number snap

Buy or make a set of 33 blank cards, the size of playing cards. Write one of the numbers from 0 to 10 on each of them so you have three sets of numbers.

Shuffle and deal the cards. The players take it in turns to turn over their top card. If two matching cards are turned over, the first player to shout "snap" wins both the piles concerned. Continue until only one player is left.

Variation

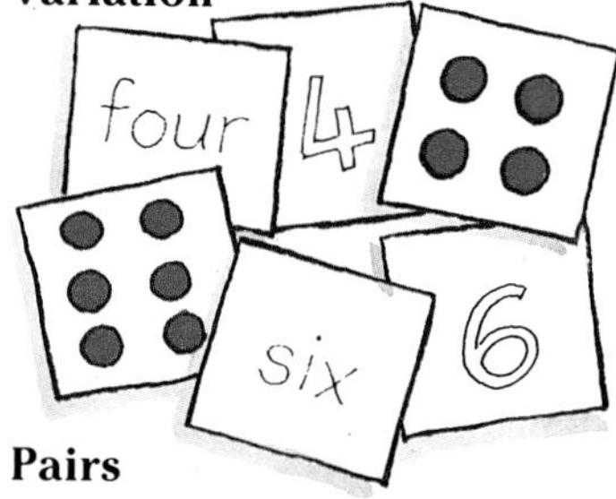

- Make a set of cards with the numbers shown in different ways, as shown on the left.

Pairs

Choose some pairs of cards from either of the sets you made for Number snap. Lay the cards out face down, either in rows or just anyhow. Take it in turns to turn over two cards. If they match, you win them. If not, turn them face down again.

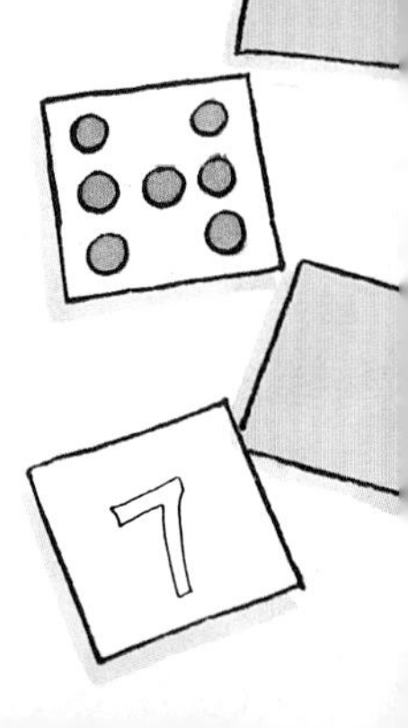

Number lotto

Make some cards with a number on one side and some lotto boards with corresponding numbers in each square. Each player has a board. They take it in turns to pick up a card and if its number appears on their board they use it to cover the number up. Continue until someone fills their board.

Variations

- One person is the "caller". She holds up the number cards and players have to shout out the name of the number in order to claim it to put on their board.

- On the back of each of your number cards, draw and colour in the corresponding number of spots. Play so that you have to match up the spots to the numbers on the boards.

Dot-to-dot pictures

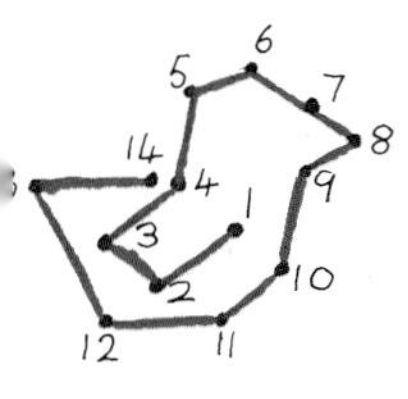

Prepare some simple pictures for your child to complete by joining up numbered dots. Drawing the dots round templates or on tracing paper over pictures in books is an easy way of doing this.

Draw a face

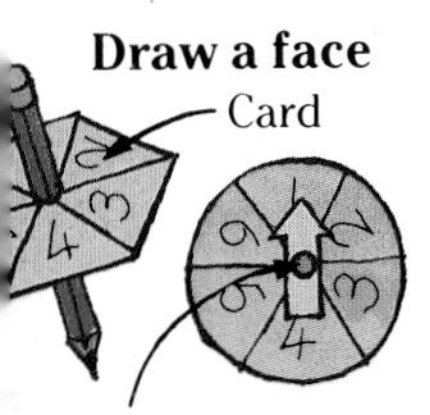

First make a number spinner or stick numbers over the dots on a dice.

Each player draws the outline of a face on a piece of paper. Players then take it in turns to spin a number and draw in a feature of the face, depending on the number spun. If you get a number you no longer need, you pass.

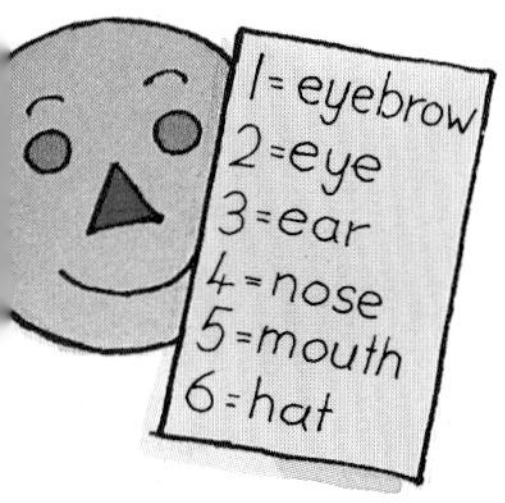

1 = eyebrow
2 = eye
3 = ear
4 = nose
5 = mouth
6 = hat

Games to buy

There are some excellent games you can buy which help children consolidate their counting skills, usually by getting them to count the spots on a dice and move a corresponding number of steps on a board.

Look out for some of the newer games besides old favourites like Ludo and Snakes and ladders.

Let's make it harder. Throw two dice and choose which one of the numbers to use.

To get more mileage from the games, you might like to try inventing variations of your own.

Number books

There are all sorts of number books on the market, including picture books illustrating the sequence of numbers, colouring books, "dot-to-dot" books, books that try to teach sums.

If a book is fun and your child enjoys it, she will probably learn something from it. If it is dull or you forget to keep your use of it casual, she probably won't. Another thing to bear in mind is that books can never replace practical experience. However many number books your child looks at, she won't learn much unless she also counts real things for real reasons.

Writing numbers

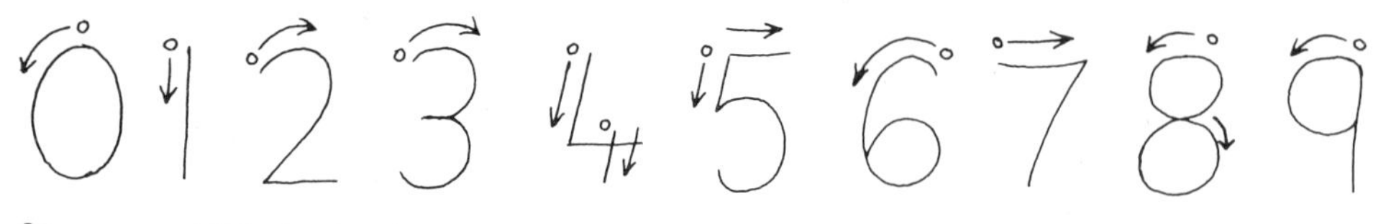

Give your child plenty of encouragement when she makes shapes resembling numbers in her drawings. If she is learning to write letters, don't forget to show her the numbers as well. Try writing numbers with a highlighter pen and see if she enjoys tracing over them with felt-tip or crayon, or do dots for her to join up. Make sure you show her from the start exactly how to form each number, as habits soon become ingrained.

Don't worry when she writes the numbers wrong, or even if she makes up her own way of writing them, but do give plenty of praise when she does something right.

Making books

Your child may enjoy making his own books with you. You can buy a scrapbook, a notebook with blank pages or a photograph album, or just punch holes in sheets of paper or card and tie them together with string, yarn or ribbon.

3
Three toys In my bed.
4
Four candles on my cake.
1 cherry left.
0 cherries left.
3 people pulled the turnip
4 people pulled the turnip
4
four
4
4

◀ Help your child to draw pictures or glue in pictures from magazines of things that are important to him and write a number caption underneath.

◀ It is a good idea to include zero in your book. Many books start at 1 but children need to learn about 0 too.

◀ You could perhaps even try making a book to illustrate a favourite number story or rhyme.

◀ Your child might like to make a collection of a specially significant number, such as his age, written or printed in different ways.

How many in the pot?

Collect a few empty yoghurt cartons or matchboxes and make some labels showing numbers in various different ways.

Hide a few dried beans under each carton or in each box and put labels beside them indicating how many are in each one. Ask your child to read the number and then check under the carton to see if she was right.

Variations

- When your child is confident at this, give her a carton and a blank label of her own. Ask her to hide some beans under the carton and write a label for you to read.
- Use just one carton and five beans. Hide some of the beans under the carton and leave the rest on view. Ask your child to write a label indicating how many beans she thinks are hidden. This will not only give her practice in writing numbers but also start her thinking about subtraction.

Playing shop

This gives excellent opportunities for counting both objects and money. To start with, give your child only coins to the value of one and have nothing in the shop costing over 10 units. Later, she will be able to start adding up different amounts and subtracting to find the right change, and will gradually learn to recognize the value of the different coins. Put price labels on the goods for sale.

Ordinal and cardinal number

When you count things, such as forks, to make sure there is one for everyone, you are using cardinal numbers. It does not matter what order you count the forks in; you just want to know how many there are.

You use ordinal numbers when you are interested in where something comes in order, for example, is Sally the first, second or third child? is the picture of the dog on page one, two or three?

Young children often get these two kinds of thinking confused. Try putting five buttons in a row and counting them with your child. Then ask her to give you three buttons. She may just give you one button, the third in the row. This shows that she thinks you were naming the buttons when you said, "1, 2, 3...". Or she may pass you three buttons but think that it matters which three (the ones you "named" 1, 2, 3).

Children eventually grow out of these confusions. Just continue using numbers in a natural way in everyday life, in their ordinal as well as cardinal sense.

5
Where's number 17?
That's the second apple you've had today.

Which pot is it in?

It's under the second pot from this end. Can you find it?

Put about seven empty yoghurt pots upside down in a row. Then, while your child closes his eyes, hide a coin under one of them. See if he can find it from your description. When it is his turn to hide the coin, he may not be able to describe its position so you may need to put questions to him, for example: "Is it under the first pot?"

Variation

- For older children, use "left" and "right", for example: the third pot on the right.

SHAPE AND SPACE

Even babies can learn about shape and space through play.

Shape and space are as much a part of mathematics as number and are just other names for geometry, which is taught at secondary school level.

We use our awareness of flat (two dimensional) and solid (three dimensional) space all day long without even realizing it and babies start to develop a feel for it even before they become aware of numbers.

Given as many opportunities for the right sort of play as possible, young children will learn a tremendous amount about shape and space by themselves, even before they go to school. The activities suggested on the next few pages will all help to increase your child's familiarity with shapes and their properties, and with the vocabulary of shape.

Physical play

Children get a lot of their early ideas about shape and space by using their own bodies to explore space: crawling through tunnels, fitting into large cardboard boxes, hiding in dens, sliding, swinging, balancing and climbing.

Using position words

Use position and direction words ("inside", "on top of", "forwards", "left", "right" and so on) in your normal conversations with your child and look out for opportunities to draw attention to them when you are reading books together.

Who's that hiding under the hedge?

Robot treasure hunt

Go straight ahead for four paces. Now turn right. Step over the box. Go through the tunnel.

Your child can only do exactly what you instruct him to in order to reach the "treasure".

Simon says

This is another way of practising position words ("put your hands *behind* your back"; "put your arms *above* your head") and also allows you to introduce some shape words ("stretch up *tall*"; "lie *flat* on the floor"). Remember, if anyone follows an instruction that is not prefaced by "Simon says", they are out.

Toys

There are many toys which can help to develop awareness of shape.

Blocks

By playing with blocks, your child will learn, among other things:

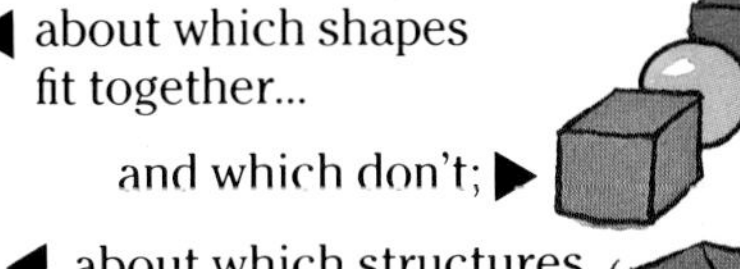

◀ about which shapes fit together...

and which don't; ▶

◀ about which structures are stable...

and which aren't; ▶

◀ about balancing.

You can add a collection of blocks with cutouts of wood sanded down. Toy animals, people and cars to go with the blocks will help to give ideas for building and encourage experimentation.

Construction kits

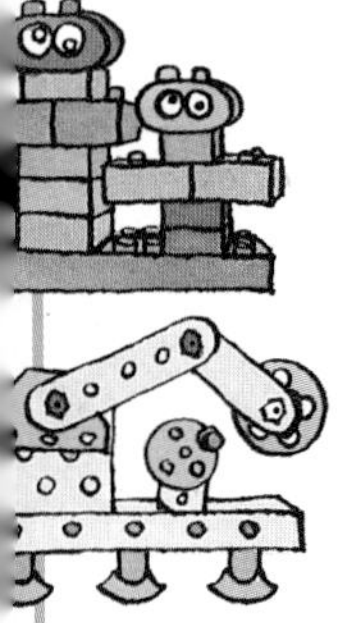

These range from fairly simple sets with large pieces for very young children to complex, technical systems for older children. Some kits consist of pieces which clip together, others include tools and things like nuts and bolts for fixing.

Blocks and construction kits have traditionally been thought of as boys' toys. Be sure to encourage girls to play with them too so they don't miss out on important mathematical experience.

Shape sorters and posting boxes

There are many different types of these, varying in difficulty. For a very simple first shape sorter, cut a round hole in the lid of a shoe box just the right size for a small ball, and a square hole for a block. Challenge an older child who manages a sorter easily to do it with her eyes shut.

Jigsaws

Start off with inset puzzles for very young children, then go on to ordinary jigsaws with very few pieces and gradually increase the level of difficulty.

An older child who can do a particular jigsaw easily might like to try doing it with the picture turned top to bottom.

Using toys with your child

Always check that toys of the sort mentioned above are well made so that parts designed to fit together do so exactly.

Remember to allow your child to play with the toys in her own way, experimenting and making discoveries for herself.

What do we need next? A bit of green for the tree, I think. Now, where can that be?

Give help and encouragement when needed but resist any temptation to take over and do things for her.

That's a steep slope. What happens if you make it less steep? Will your car still go down it?

Extend your child's vocabulary of shape by talking with him about what he is doing and gently introduce new ideas. Don't insist, however, if your child prefers to play on his own.

Learning shape names

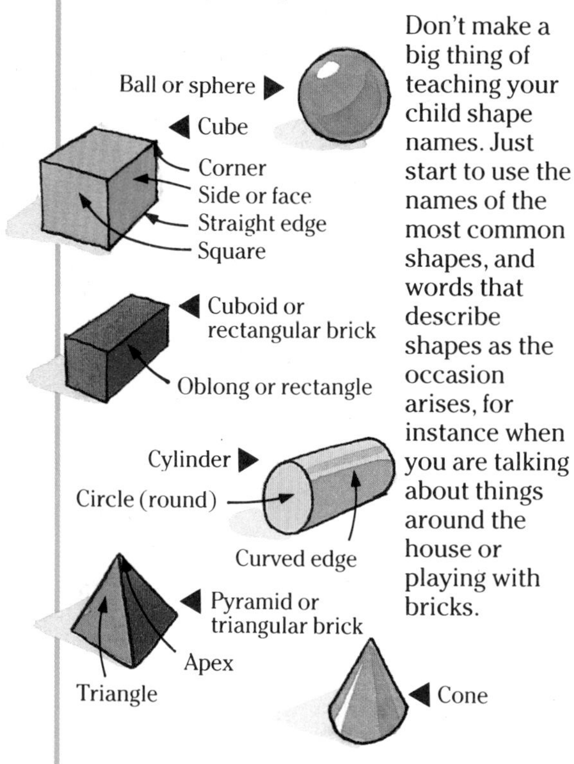

Don't make a big thing of teaching your child shape names. Just start to use the names of the most common shapes, and words that describe shapes as the occasion arises, for instance when you are talking about things around the house or playing with bricks.

Drawing, cutting and sticking

Cut out shapes from card for your child to draw round or help her draw round real objects. She might also like cutting shapes out of paper, perhaps from pictures in magazines and glueing them on a piece of paper.

Playdough

Make and cut out different shapes from playdough.

Printing

Try printing with blocks, vegetables or other things. Use powder paints or liquid poster paints and print onto thick, non-shiny paper with a wad of newspaper underneath. Use a brush to put the paint on the printing object.

Shapes lotto

Make some boards with shapes drawn on them and cut some shapes out of card to match. One person holds up each shape in turn for the players to claim. Continue until someone fills their board.

Variation

- Play with blocks instead of card shapes, matching one face of the block to the shapes on the boards.

Build a house

Make some boards with a picture of a house drawn on them as shown on the right. Then cut out some shapes identical to those used in the house picture. Make a spinner as shown.

Players take it in turn to spin and pick up a corresponding shape to put on their board. If you get a shape you don't need, you pass.

Shapes I-spy

Instead of first letter sounds, use shapes, for example, "I spy, with my little eye, a circle." There will be no shortage of shapes around the home.

Circles: clocks, buttons, plates, records, rings, pan lids, coins, the moon. Squares: floor tiles, paper napkins, paper towels, chair seats. Rectangles: windows, doors, books, rugs, pictures, envelopes. Triangles: sandwiches, roof tops, pushchair frames.

Feely bag game

Get together two sets of blocks of identical shape and size. Put one set out on the table and, without your child seeing, put one of the blocks from the other set into a bag. Your child feels the block in the bag and has to point to the one on the table which is like it.

Variations

It's like the square one. No, I mean the one with all squares, not oblongs.
Oh, you mean the cube.

- Get her to tell you which is the matching block on the table without pointing.
- Let her put a block in the bag for you to feel. Your description will give her clues about words she can use when it's her turn.

It's sharp. I think it's the shape with a triangle at both ends.

- You can also play just by putting a block in the bag without having a matching set to compare with. Try using other objects too, such as a marble, pencil, matchbox or dice and guessing what they are.

Making things from junk

Making things from junk helps to develop an awareness of shapes and their properties. Provide sticky tape or glue and collect boxes (cereal and tea packets, shoe and match boxes, juice containers, egg boxes), tubes (kitchen and toilet roll, dishwashing liquid bottles, plastic herb pots) and tubs (margarine, ice-cream and yoghurt cartons).

Symmetry

Many children can begin to recognize symmetry at a surprisingly young age.

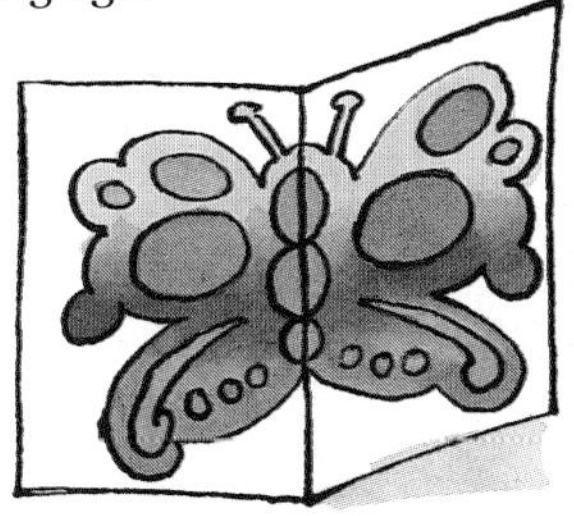

Try painting a symmetrical picture by folding a piece of paper in half, then opening it up and putting fairly runny paint on one half near the fold. Fold the clean half over and press hard all over before opening it up again.

Your child might also appreciate symmetrical shapes cut out of paper. Fold a piece of paper in half, cut out half of a shape along the fold, then open out.

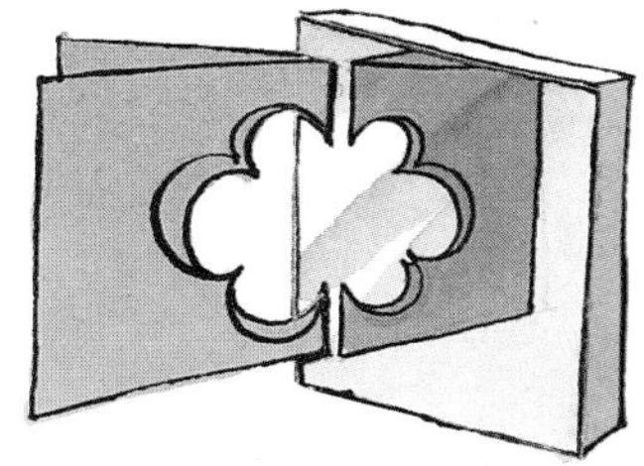

You can then fold the paper back on itself so that only half the picture shows and hold it up at right angles to a mirror to re-create the original.

Symmetry squares

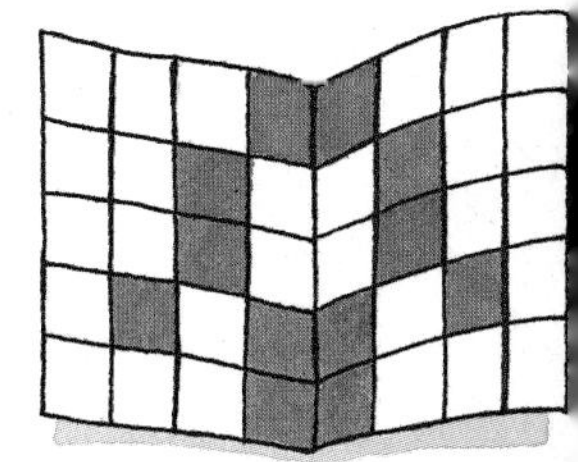

Two people can use squared paper to make a symmetrical pattern. First draw a line down the middle of the paper. Then one person colours in a square on her side and the other colours in the mirror-image square on his side. Then the second person colours in one of his squares and the first colours in the mirror-image. And so on.

SORTING THINGS OUT

... those because they're red, that because it's wood and that's blue.

Sometimes children change the way they are sorting in mid-stream.

Sorting is an everyday activity and one which children start to do quite naturally from an early age. For instance, when babies are learning to differentiate dogs from cats, they are doing a form of sorting.

Children have to be able to sort before they can learn to count in a useful way. For example, a child cannot pass you three apples unless she is able to tell the apples from the rest of the fruit.

Sorting is also important because it encourages children to think about the logical aspects of sameness and difference and how things can be categorized. This is a basic part of mathematical thinking.

Young children often sort in a way that seems odd at first. By all means ask your child what she has done but don't correct her if you think she is wrong. She may just be playing, or she may be sorting according to some criterion of her own: this may even be more logical than you think. She will learn to sort in a more conventional way as she sees how and why other people sort things in real life.

Odd one out

That one because it's got a stripe on it.

Spotting the odd one out will get your child thinking about sameness and difference. You can make the game harder as he gets older. For instance, see if he can spot the difference in textures by having three hard objects and one soft, or in materials by having three plastic things and one metal.

Be sure to acknowledge that he is right if he spots a difference that you had not noticed. You can always tell him what you were thinking of afterwards. Let him give you some things so you have a turn at spotting the odd one out too.

Sorting for real

Let your child do things like putting away the cutlery after you have washed up, putting away the shopping, sorting the washing into piles to be put away and pairing up the socks, arranging books on the bookshelf. Young ones will probably need to sort books according to size or colour but older ones might like to do it according to content, for example, all the fairy stories together, all the adventures and so on. If motivation is needed, you could try making the task into a challenge.

Bet you can't put all those toys away where they belong before I've got the supper ready.

Things to sort

To give your child the idea of sorting in play, you may need to start casually sorting things out yourself. Try sorting out all the red blocks to make a tower, sorting coins into piles of different denominations, putting all the sheep in one field, all the cows in another.

Bits and bobs box

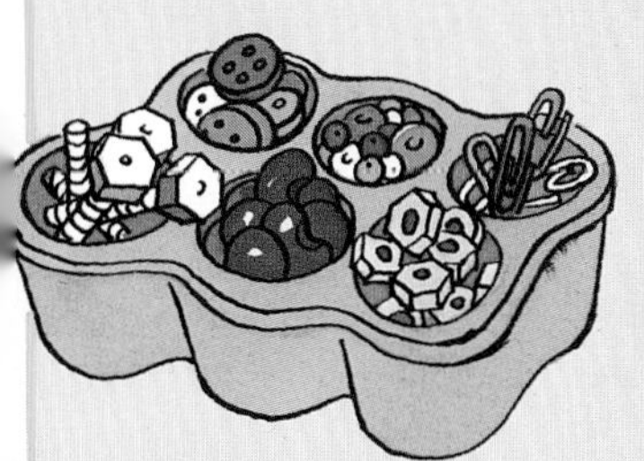

Keep a special box full of small objects like buttons, paperclips, beads, dried beans, coins, nuts and bolts. Your child may like to have an egg box or a muffin tin to sort them into.

Scrapbook

Cut out pictures from old magazines and catalogues, and save birthday and Christmas cards, so that you can make a scrapbook divided up into different categories such as animals, cars, food, babies. You could also incorporate numbers.

What's my set of shapes?

Start picking out all the blocks of a certain shape, say the "round" ones, and see how soon your child can guess what you are doing. You and your child will probably have to make your criterion quite clear. By "round" do you just mean ball shapes or are you including cylinders or shapes where at least one of the faces is a circle?

What's my set of numbers?

From a set of number cards or playing cards, start picking out all the numbers say, over 6 or which are multiples of 5.

Blocks lotto

Try making a lotto game based on the ideas that you and your child came up with when playing "What's my set?" above.

Cut out some boards and write one description in each box, as shown on the right. One person holds up a block and if it fits one of the player's descriptions, they claim it to put on their board. If a shape fits more than one player's board, the first to claim it, wins it. Continue until someone fills their board.

MAKING PATTERNS

Patterns are important in mathematics, both visual and number patterns. This is because math is all about things being orderly, regular and systematic.

If children begin to expect patterns in math, the subject becomes much easier and more friendly. They realize that it isn't about learning lots of different, unconnected facts and ideas It also makes them better able to work out facts they don't know. For instance, a child who knows the pattern of the 5 times table but can't remember what 7 times 5 is can simply count "5, 10, 15, 20…" until she reaches the seventh in the series.

Patterns all around

Draw your child's attention to patterns around the house, in fabrics, wallpaper, tiles or carpets, for example. Have a close look at the patterns together and see if you can spot where they start to repeat.

You can also encourage your child to look for patterns in nature, in cobwebs, flowers, leaves, shells, cut fruit and vegetables for instance.

Pattern rubbings

Children often enjoy looking for things to take rubbings from. Just put a piece of paper over objects with a textured surface and rub with a wax crayon.

Brickwork

Loudspeaker

Leaf

Try things like brickwork or fencing, leaves and tree trunks, embossed wallpaper, shoe soles, loudspeakers, and anything made of wicker.

Drawing, printing and making collages

Your child might enjoy drawing simple patterns, or using the printing ideas on page 12 to produce patterns, or making a collage pattern.

In each case, start off by suggesting that he does one row of pattern, then you copy it underneath. Then you do a row for him to copy, and so on.

If you cover a whole sheet of paper with a pattern, you could then use it as wrapping paper.

Paper cutting

Fold a piece of paper into a concertina or fold it in half three times and experiment with your child to see what patterns you can make by snipping bits from the sides and then opening out.

Give your child plenty of time to play around with this activity. Later, she may welcome challenges such as making a string of people.

Threading

Playing with threading toys such as cotton reels or blocks, or making bead necklaces, can be an excellent way of developing an awareness of pattern.

Ask your child to start off a necklace by threading two or three beads on a string. Continue by repeating his pattern over and over again. When he realizes what you are doing, involve him by getting him to tell you what colours you need to thread next, or let him take over if he wants to.

Variations

- Make the activity more complicated by starting off with a basic pattern and then producing it in mirror image. Continue to repeat the pattern and its mirror image until the string is full.
- Try threading half a necklace and then producing a mirror image of the whole sequence for the second half.

If your child makes a mistake, it may be best not to point it out immediately but to wait and see if he notices it himself. If he doesn't, suggest at the end that you both look through the necklace to check you have done it right.

You can also encourage your child to develop the habit of checking by telling him you are going to make a deliberate mistake and challenging him to spot it.

Music

Copy a rhythm your child has produced, perhaps on one of her instruments, either by clapping your hands or repeating on another instrument. Then make a rhythm for her to imitate.

Variation

- Take it in turns to produce rhythms for each other to copy by hitting two different household "instruments" with something.

Blocks

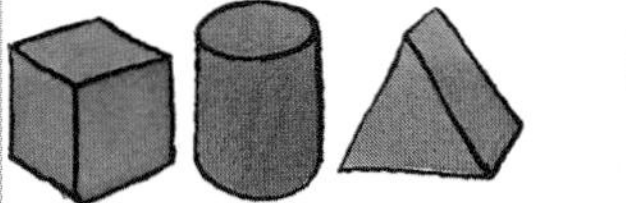

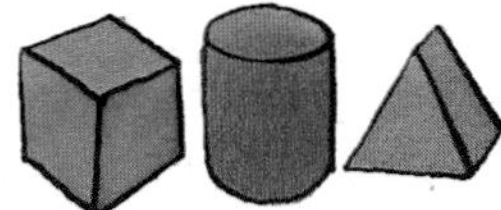

Start casually making a line of different-shaped blocks and see if your child can continue the pattern.

Write down the numbers 1 to 10 on pieces of paper. Get your child to choose one of the numbers and take that many blocks of the same shape and size from her set.

Can she use the blocks to build two towers exactly the same height? This is a way of introducing the idea of dividing by 2, and odd and even numbers.

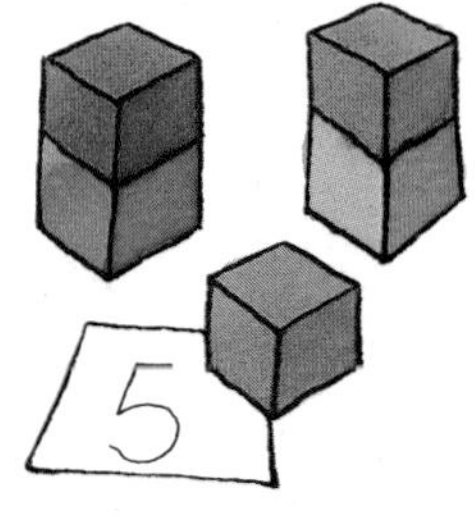

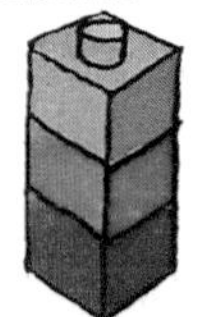

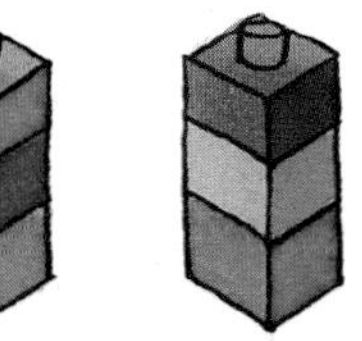

How many completely different towers can you make using three different-coloured blocks? If you want to increase the permutations, try allowing each of the colours in turn to stay in the same position for two of the towers.

PUTTING THINGS IN ORDER

Ordering, or sequencing, is very important in everyday life as well as in math. You need to follow the correct sequence of actions to do things like drive a car or bake a cake. You use alphabetical order to look up a word in a dictionary and you use number order all the time.

Your child can learn a lot about order simply by watching you and helping with everyday tasks; from listening to stories where order is important; and, most important of all, through conversation involving order words such as “before”, “then”, “after”, “first”, “second”, “third”.

Toys

There are several toys which are excellent for introducing the idea of ordering to even very young children. Remember, though, that they are toys and let your child play with them in her own way.

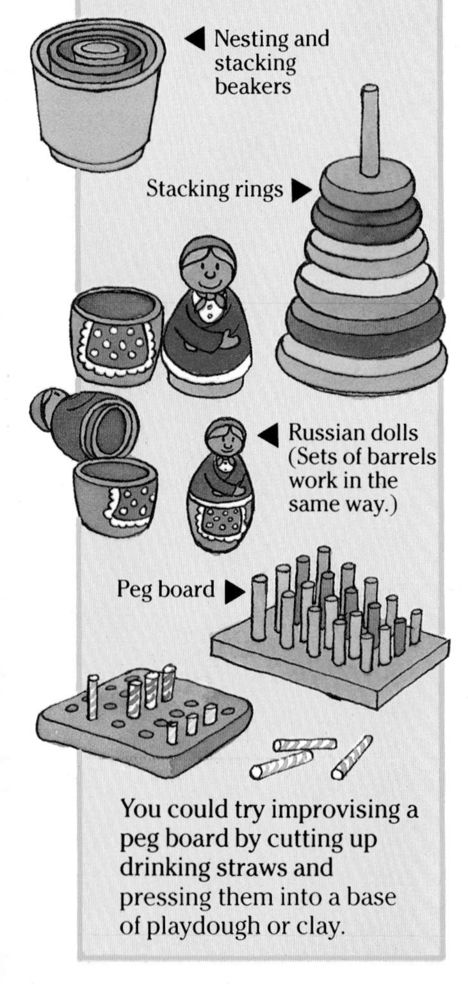

You could try improvising a peg board by cutting up drinking straws and pressing them into a base of playdough or clay.

Ordering games

Which card is missing?

Put in order a set of number cards, playing cards or cards with spots drawn on them, say 0 to 5 or 0 to 10. Then, while your child closes her eyes, hide one of the cards and close up the gap. Which one is missing?

Variations

- Swap two cards around instead of hiding one.
- Spread the cards out anyhow, then hide one.

Where does it go?

Put a set of number cards in a pile. One by one take the card from the top, judge whereabouts in the sequence it belongs and position it on the table accordingly. At the end, all the cards should be in the correct order, and neither overlapping nor too spread out. Once a card is placed, you may not move it.

Walk the number line*

Chalk a line outdoors, long enough to include all the numbers 0 to 10 one step apart; make a mark where each number will come but don’t actually write them in. Shuffle a set of number cards.

Take it in turns to take a number card and walk along the line from the start, stepping on the markers and counting out loud as you do so. When you get to the correct spot for your number, put the card on the ground. When all the cards are in position, turn them face down. Ask your child to walk slowly along the line from the beginning until you shout “stop”. Then see if she can tell you which number she is on. Check by turning the card over.

For more about number lines, see pages 22 and 37.

Robots

Fasten your shoelace. Now put on your sock . . . Oops!

Take it in turns to give each other a sequence of precise instructions for performing a simple everyday activity such as getting a drink, emptying a wastepaper basket or getting dressed.

Noises

Get your child to close his eyes while you make a series of about three noises. See if he can tell you what order they came in.

Three in a row

Make two or three sets of number cards, 0 to 10. Shuffle and deal six cards to each player. (Two or three can play.) Put the rest of the cards in a pile face down, then turn over the top card.

The aim of the game is to make up a set of three numbers in sequence by taking turns to pick up a card and throw one away. First to get a set of three wins.

Tricksy

Make two or three sets of number cards, 0 to 10. Shuffle and deal six cards to each player.

Everyone takes turns to put down one card. Whoever puts down the highest card wins that "trick" (the cards on the table). She puts those cards on one side and starts off the next round. At the end of the six rounds, the winner is the person with the most tricks.

Bigger and smaller

Write the numbers 0 to 10 on a sheet of paper. Make a dice by glueing "bigger" on three sides of a block and "smaller" on the other three. Put a toy car on 5.

Take turns to throw the dice and move the car to a bigger or a smaller number, depending on your throw. Cross the number out and leave the car there for the next person. You cannot move to numbers which have been crossed out so sometimes you cannot move. The person who crosses out the last number wins.

Variation

- Instead of writing a row of numbers, draw a row of 11 circles. When a player lands in a box, they write in the correct number.

ADDING AND SUBTRACTING

Children need a lot of practical experience of adding and subtracting.

As children get older, they meet situations where they need to combine two different sets of things (add) or to split a set up, to count differences and to compare (subtract).

For a long time, they need to do their adding and subtracting in a very practical way, for a real purpose, with real objects in front of them. It is vitally important not to rush children through this stage and on to doing abstract calculations. They need to add three things to two things to make five many, many times before they realize that 3 plus 2 always equals 5.

However, there are ways in which you can help them move towards that realization and some of these are explained on the next few pages.

Getting started

Continue with the activities suggested on pages 4-9 for developing early number skills and make use of any opportunities for adding and subtracting in daily life as your child becomes more competent.

Mealtimes can be excellent for doing some casual calculations so long as your child enjoys it. Remember to stick to small numbers at first and try asking questions along the following lines.

How many more plates do you need?
Who's got more? ... How many more?
Who's got fewer?
How many grapes have you got? ... If I give you 2 more, how many will you have then?
You had 5 and now you've eaten 1; how many have you got left?
You've had 2 already and you want some more! I'll give you "none" more. How many will you have had then?
5 for you and 5 for me makes 10.

Remember to include nothing in your calculations.

Giving two people the same number of things will not only introduce your child to dividing by 2 but can also be a way of learning "doubles".

Stages in learning to add and subtract

Don't be concerned about how your child arrives at the right answer to a question. When children are learning to add up, for instance, they may use various strategies. If you ask a child who has six crackers to take three more and tell you how many he has altogether, he may count all the crackers from the beginning ("1, 2, 3...etc.); he may "count on" from the total of one group of crackers ("6, 7, 8, 9") or he may not count at all but know from experience that 6 and 3 always make 9.

Writing sums down

Don't try to rush your child through the stages described above or think that you may clarify things for him by writing down the sum, for example, 3 + 2 = 5. This will be too abstract for him and is not a method that is generally used even in schools until a child has a good understanding of numbers.

First of all children are encouraged to experiment with their own ways of writing or drawing sums. This is the only way to be sure that they understand what they are doing.

Later, as they compare their methods with other children's and become concerned to find the quickest way of doing a calculation, the teacher may introduce more conventional forms of laying out sums.

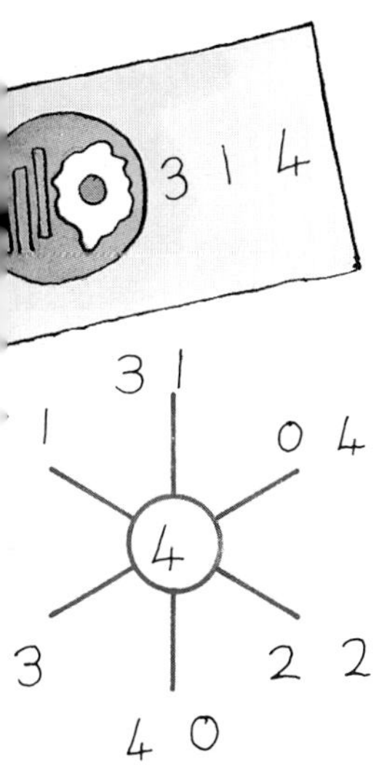

Children's early sums. Ways of making 4.

Using your fingers

3 and how many makes 10?
3 and 7 makes 10.

Using your fingers can be very helpful, especially for learning the pairs of numbers which make 10. These are important "number bonds" (see page 23) and it is worth spending some time on them.

First, though, make sure your child is quite certain how many fingers everybody has. Count each other's and see if he still knows how many when you are wearing gloves or hiding your hands behind your back.

What words to use?

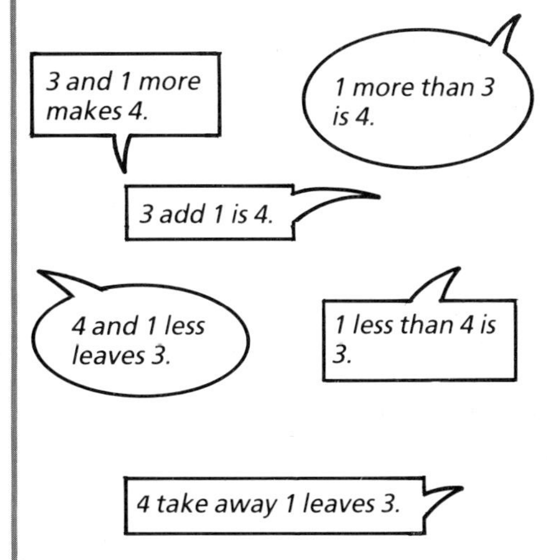

3 and 1 more makes 4.
1 more than 3 is 4.
3 add 1 is 4.
4 and 1 less leaves 3.
1 less than 4 is 3.
4 take away 1 leaves 3.

When you are adding and subtracting with your child, use whichever form of words your child appears to understand and which comes naturally to you. It is a good idea to vary the way you express the same thing to be sure that she understands the principles involved thoroughly.

Number lines

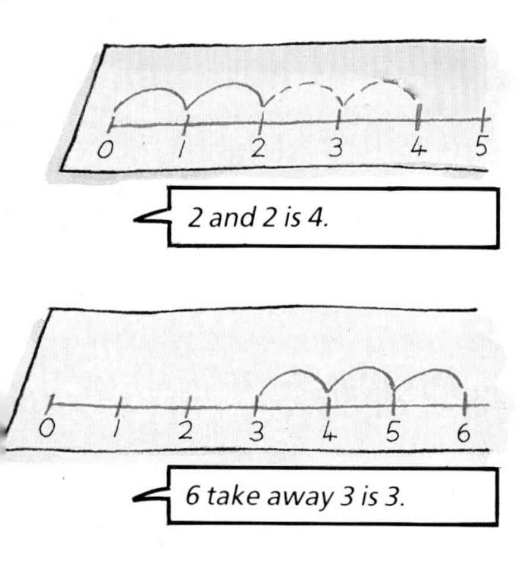

Number lines are often used in schools to help children build up a mental image of numbers in order and to help with adding and subtracting. Children can count along the line, forwards and backwards, draw steps or move a counter along it, or even walk along a floor number line, one step per number.

When children first start "counting on", say when they are playing a board game, they often make the mistake of counting the number that they are already on. The way the number line is drawn emphasizes the space between the numbers and so encourages children to count accurately by taking a step forwards rather than a step on the spot.

As children get older, the lines can be extended to include bigger numbers and may also be used to show negative numbers, fractions and decimals.

Number line game

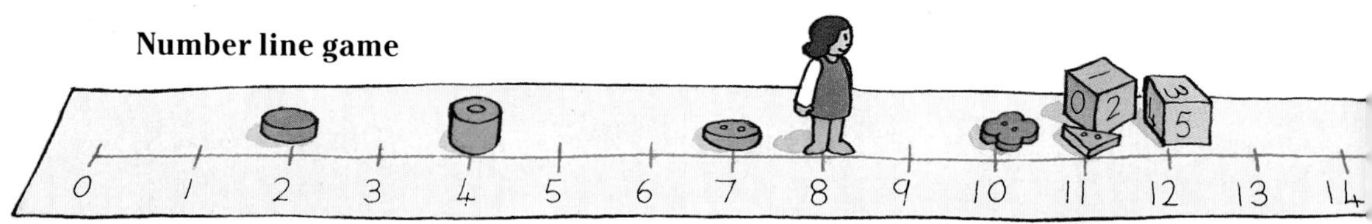

Draw a line marked out with the numbers 0 to 20 a few centimetres apart either on a big sheet of paper or outdoors. Put a bead or button at ten of the numbers (but not the 0 or 20). If possible make two dice, each numbered from 0 to 5, or use two ordinary dice. Put a miniature doll, toy car or counter for each player at 0.

Take it in turns to throw both dice and choose one of the numbers to move for. If you land on a number with a bead or button, you can pick it up. When you reach 20, turn round and come back again. You must throw the exact number to land back on 0. The winner is the person with the most beads.

Encourage your child to pay attention to the numbers by doing so yourself.

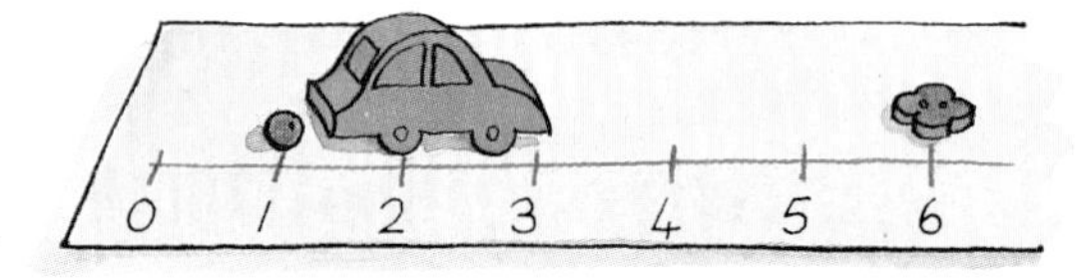

Variations

- To make the game non-competitive, you could pool the beads you pick up and perhaps play with only one car between you.
- Both start off at number 10. After each throw, choose which direction to go in as well as which number to move for. Carry on until all the beads are collected.

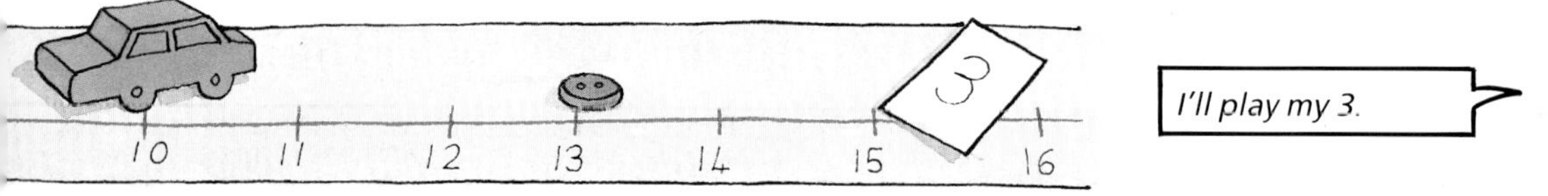

- To get your child thinking about the numbers, try playing with number cards 0-10 instead of dice. Shuffle the cards and share them out at the start of the game. When it is your turn, you choose which of your cards to put down. Then move that number of steps. Keep going until all the cards are used up.

Number bonds

You can help to provide your child with a really sound basis for doing calculations by helping her to master number bonds.

These are simple number facts, such as 2 and 2 makes 4, and a lot of work is done on them in schools.

The chart on the right shows all the addition bonds your child needs to know. To find 3 plus 5, look for the square where the row starting with 3 meets the column headed 5. Or, look for where the column headed 3 meets the row starting with 5.

The number of bonds to be learned may seem daunting. Bear in mind, though, that each bond is shown twice (3 plus 5 is the same as 5 plus 3).

Never try to teach your child the bonds directly from the chart. It is only intended as a checklist. You may like to cross off the bonds as she learns them. When you are playing one of the games suggested on the next two pages, tick off each bond she knows. Then, if she remembers the bond another day, you could cross it right out.

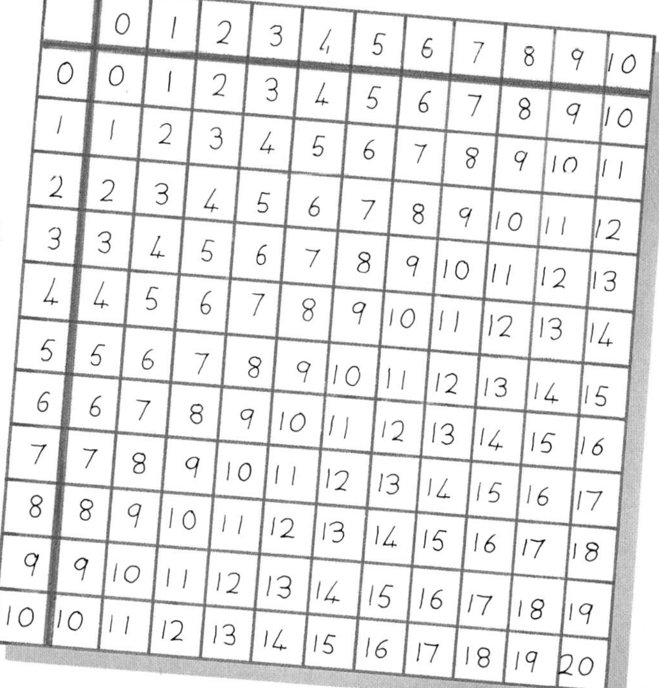

	0	1	2	3	4	5	6	7	8	9	10
0	0	1	2	3	4	5	6	7	8	9	10
1	1	2	3	4	5	6	7	8	9	10	11
2	2	3	4	5	6	7	8	9	10	11	12
3	3	4	5	6	7	8	9	10	11	12	13
4	4	5	6	7	8	9	10	11	12	13	14
5	5	6	7	8	9	10	11	12	13	14	15
6	6	7	8	9	10	11	12	13	14	15	16
7	7	8	9	10	11	12	13	14	15	16	17
8	8	9	10	11	12	13	14	15	16	17	18
9	9	10	11	12	13	14	15	16	17	18	19
10	10	11	12	13	14	15	16	17	18	19	20

Bonds and subtraction

Knowing the addition bonds will also help your child with subtraction.

If she knows that 3 and 7 makes 10, she should also realize in time that 10 can be split up again into 3 and 7. And so she can work out that taking 3 from 10 must leave 7, and taking 7 from 10 must leave 3.

Visualizing numbers

I'm holding up 6 fingers. How many are bent down?

It is very useful to be able to visualize numbers, whether by imagining fingers held up or a number line of some sort.

5 and 7? Well, 5 and 5 is 10. Add 2 more makes 12.

If your child can see numbers in her mind and re-arrange them, she will be better able to make use of what she already knows to find a way of working out number bonds she does not know. It will also stand her in good stead for doing more complicated calculations later on.

Games to help teach number bonds*

Think of a number

I'm thinking of a number which is 3 more than 5. Do you know what it is?

Your child will probably respond more enthusiastically to this sort of challenge than to a straight question: "What's 3 and 5?" She can use a number line to check her answer.

How many have disappeared?

Ask your child to count how many strawberries are in your bowl and then shut her eyes while you eat some? Can she work out how many you have eaten?

If she finds this difficult, try just hiding some, so that she can check her answer by counting when they re-appear.

There were 5 people in the field. How many are hiding?

She might also like to play this with toy animals or dolls.

Hoopla

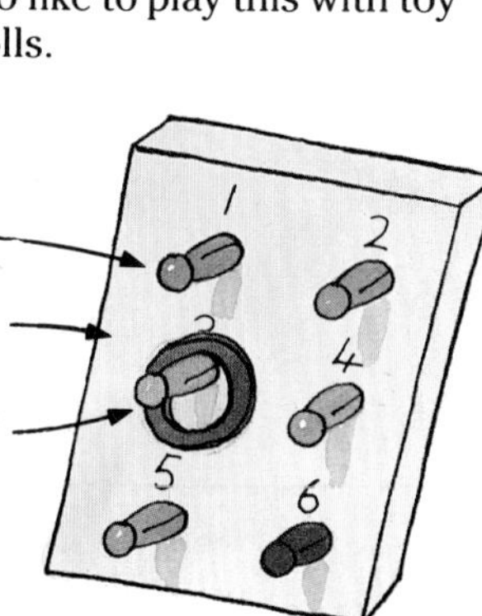

Hoopla gives practice at adding up and you can vary the points awarded to suit the stage your child is at.

Darts

This is another good way of practising adding up or subtracting, with the help of counters if necessary. Be sure to use a children's safety-darts set.

Skittles

Working out how many skittles have been knocked down and how many are still standing can give useful practice at number bonds.

Dominoes

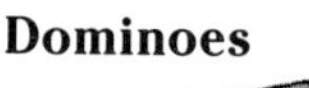

Instead of matching up like with like, play so that matching domino halves must add up to 6.

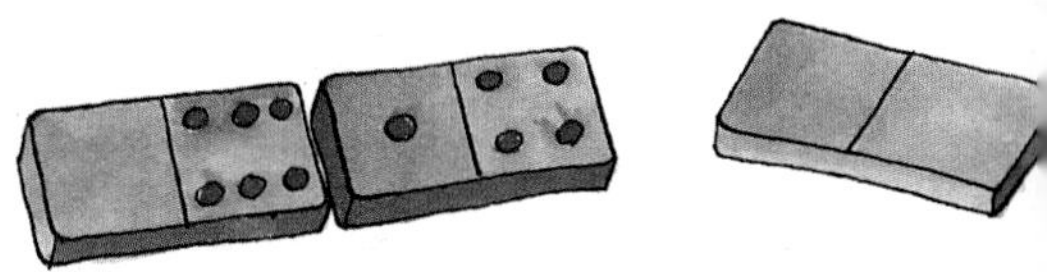

Halves add up to 7. Blanks are left over.

Halves add up to 4. 5s and 6s are left over.

You can choose other totals for the halves to add up to but you will then have some dominoes left over at the end of the game unless you remove them first.

*See also the number line game on page 22.

Dice games

Play board games using two dice and adding the two numbers together.

If you are playing on a numbered board and your child throws a 10, encourage her to predict what number she will land on before she moves her counter.

Once she knows how to add 10, adding 9 is easy too. Suggest the trick of adding 10, then taking off 1.

Catch

Pair already won.

Spare card allows players to work out which cards their opponent has.

Use one suit from a pack of playing cards. The ace is worth 1, the jack 11, the queen and king are both worth 12 and the rest are their face value. Shuffle the cards and deal six to each player. Put the last card face up on the table.

Decide who is to be the "thrower" and who the "catcher". (It is sensible for you to be the thrower at first.) The thrower puts down one card, face up. The catcher has to try and put down a card that will make it up to 13. If he does, he wins both cards. If he can't, the thrower wins both cards. The thrower continues throwing a card and the catcher tries to catch it until all the cards are used up. The one with the most pairs of cards at the end wins.

Variations

- Put the spare card face down.
- Change the total to a slightly higher or lower number.

Snap I

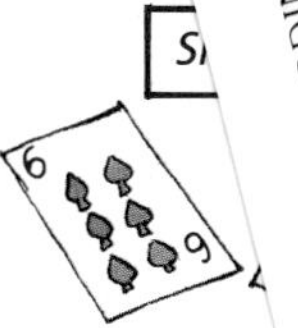

Remove the 10s, jacks, queens and kings from a pack of playing cards.
Shuffle and deal the rest. Each player turns over their top card, either in turn or simultaneously. If two cards are turned over which add up to 10, the first player to shout "snap" wins the two piles of cards concerned. Continue until only one player is left.

Snap II

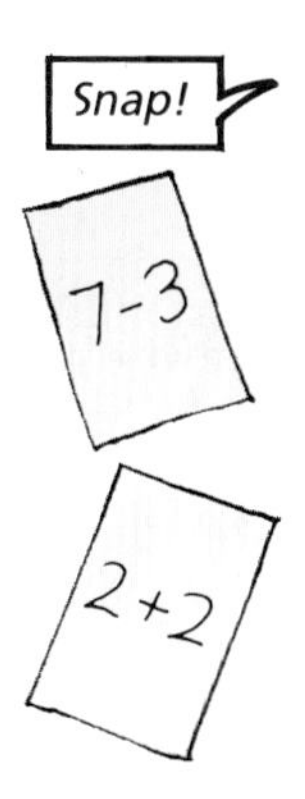

Buy or make a set of 42 blank cards, the size of playing cards. On 21 of the cards, write one each of the numbers from 0 to 20. On the other 21, write addition and subtraction bonds. Shuffle and deal the cards.

Everyone turns over their top card. If a number bond card corresponds with another number bond card or with an ordinary number card, players shout "snap".

Variations

- Abandon bonds your child gets to know well and add others that need practising.
- You can make a game for more people or with higher chances of getting snaps by adding more cards to the pack (equal quantities of bonds and "answers").

Pairs I

Remove the 10s, jacks, queens and kings from a pack of cards. Lay out the rest face down. Take it in turns to turn over two cards. If they add up to 10, you win them. If not, turn them face down again.

Pairs II

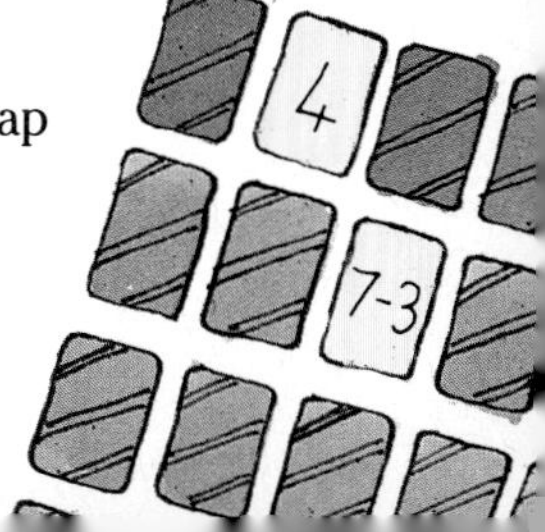

Choose some pairs of cards from the set you made for Snap II. Play as for Pairs I, only trying to turn over bonds and answers.

ng bigger numbers

Your child will probably enjoy reciting bigger numbers, will gradually learn to count to higher numbers and can learn to read and write them in the same way as he did smaller numbers. (You can adapt the activities suggested on pages 5-9.)

See if your child can say big numbers out in full ("six hundred and forty two" instead of "six four two") when you spot house numbers or car number plates.

Place value

To do calculations using bigger numbers correctly, children need to understand "place value". This means understanding that a number has a different size depending on where it is placed in a group of numbers. For example, the two symbols used in 33 are the same but have a different value because of their place.

A lot of work is done on this in schools. Children will build towers of bricks, books or coins in groups of 10, so that they learn, for instance, that 23 is two lots of 10 and three 1s.

Very young children need to develop their awareness of position (where something is) and orientation (which way round it is) before they can even begin to grasp place value. They often write letters and numbers back to front and get muddled about what order they go in. Just draw your child's attention to the right way of writing numbers, without feeling you have to correct every mistake.

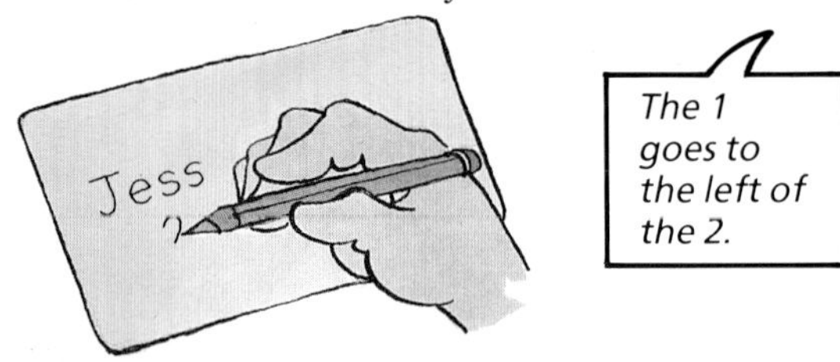

The 1 goes to the left of the 2.

Teaching him about "right" and "left" will also be useful when you begin to talk about the position of numbers.

Snap and Pairs

Make two identical sets of cards showing letters and numbers your child regularly gets back to front or in the wrong order and use them to play Snap and Pairs.

Tricksy variation

Make a set of number cards showing numbers which your child regularly muddles up and use them to play a game of Tricksy (see page 19).

If you have a number line or tape measure handy, your child can use it to check the numbers if necessary.

5 : 23
That 3 will change to a 4 soon.

Number displays

Look at digital clocks and gas pumps with your child. Talk about which part of the number changes quickest, the part on the left or the part on the right, and see if she can predict what number will come up next.

Calculators are also useful for learning about place value.*

Money

Learning the value of different coins and doing sums with money can help children to understand about place value, besides being of practical use in itself. Many children will be quite enthusiastic, especially if organizing pocket money is involved. Using money is also a good introduction to decimals.**

*See pages 40-42.
**For more about decimals, see page 42.

MEASURING

When they go to school, children will learn about all kinds of measurement but they can begin to develop a feel for them long before this stage through play and talk at home.

The first and most important idea that children need to understand about measures is what they are: they need to know what length is, for instance, and that it is different from volume or weight. This can be difficult for them for various reasons. Everyday speech does not always make the distinctions between different measures clear (see below); some measures, such as weight or time, cannot be observed directly; in addition, children have to get to grips with the difficult concept of conservation when they are dealing with measures just as they did when learning about numbers (see page 6).

Involving children with measuring at home is an ideal way of ensuring they realize that it has a practical purpose and is not just an academic exercise.

Involve children with measuring at home.

Using measurement words

Bring measurement words into your conversation from the time your child is very young. He will begin to absorb some ideas about measurement just from hearing you thinking aloud.

Look, your new trousers are longer than your old ones.
Please could you give me three of the heaviest potatoes.
Is the cake big enough for all of us?

Which container would do to put this drop of sauce in?
Will that be enough paper, do you think?

Children often have difficulty grasping the meaning of measurement words because of the imprecise way in which they are used. "What a big girl you're getting" does not really make clear whether the child is taller (her length is greater), is taller and wider (her volume), heavier (her weight) or just that she is older. Without being too pedantic, it is worth trying to use measurement words accurately.

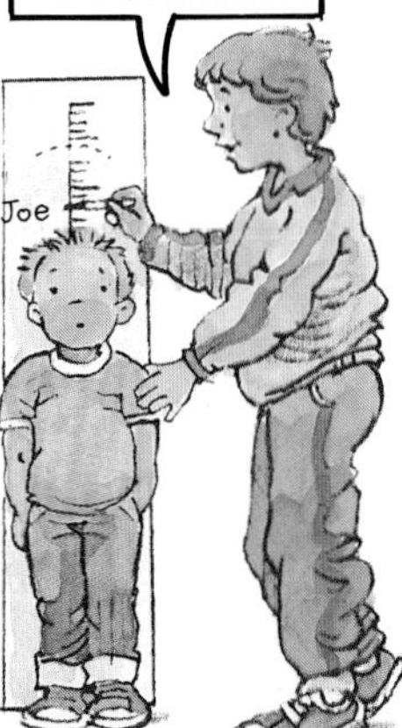

Joe is taller than you but I think you are bigger than he is overall.
Joe

Making comparisons

What a huge spider!
It is big but it's much smaller than we are, isn't it?

Young children's interest in measurement often arises from wanting to make comparisons. Make the most of this, not only to help them with their measuring but also to point out that measures are all relative. Remember to talk about things being "the same as" each other too.

Putting things in order

Making a game of putting objects in order according to some aspect of measurement may help to stimulate your child's interest in measuring. You can order according to height (use jars); length (use ribbons); or size (use jar and bottle tops, hand or footprints, balls of playdough or pastry, bowls or saucepans).

Put the objects in order. Ask your child to look at them and then close his eyes while you remove one of the objects and close up the gap in the line. Give him the object and see if he can find the right place for it in the line.

Variations

- Swap two objects around instead of removing one.
- Challenge your child to put the objects in order himself.

Units of measurement

Children can help with measuring from an early age if you use informal measures such as spoonfuls or footsteps. The formal units of measurement such as inches or pounds are difficult for children and are not introduced even in school until they have had a lot of experience of measuring in informal units such as pencils or paperclips.*

Oh dear. I wanted a bigger bottle. I wanted a gallon one and this only holds one quart.

As your child gets older, you can start mentioning the standard units when the occasion arises.*

Estimating

I think that's about enough for the parcel, do you?

A lot of everyday measuring is really estimating. Help your child develop the ability to judge measures by eye or feel (in the case of weight) by asking for his opinion when you are estimating.

That's exactly right to go round your collar.

Make sure she knows when an estimate is sufficient and when, on the other hand, a very accurate measurement is needed. Get into the habit of estimating measures together before you check them.

** Don't worry if you find yourself sometimes using imperial measures when talking to your child. While these continue to be used in Britain, children need to know something about them.*

Conservation

It's not fair. She's got more than I have.

It is not immediately obvious to young children that if a piece of string is wound into a spiral its length does not change; that there is the same amount of wrapping paper, whether the sheet is still in one piece or has been cut in four; that the cheese you have just grated will weigh the same as the original lump; or that there may be as much liquid in a short fat container as there is in a tall, narrow one.

As with conservation of number, children gradually acquire the idea of conservation of measures as they gain experience.

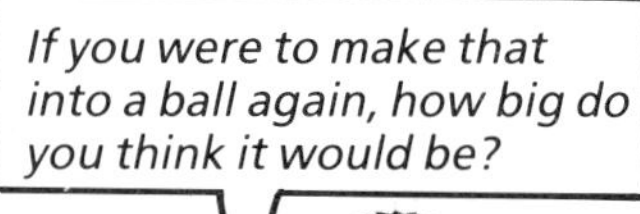

If you were to make that into a ball again, how big do you think it would be?

If you want to see if your child is getting the idea, try an experiment next time you play with dough. Suggest that he cuts the dough in half and makes two balls from it. Get him to squash one of them flat, then ask him the question above.

Length

Playing

Keep old bits of ribbon, lace, string and yarn for your child to play with. He can use them to make belts for dolls, to tie up "parcels", to measure against each other or against other things, or just to cut up to see how they get shorter.

Early measuring

Many children will spontaneously measure one thing against another to find out which is longer, or try to find out how many of one thing they can fit alongside another.

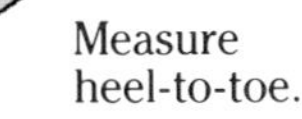

Measure heel-to-toe.

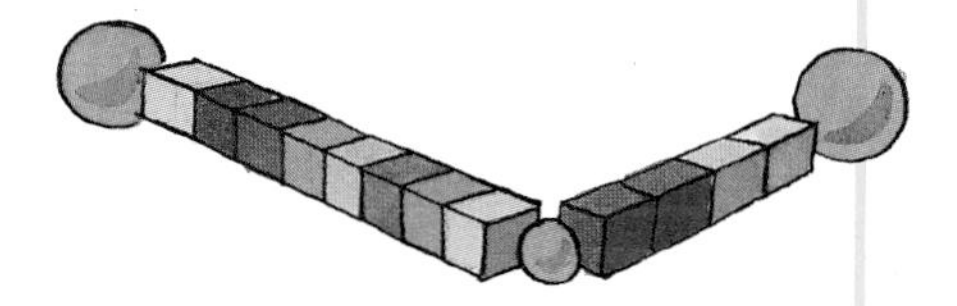

You can give the activity more purpose by estimating who has thrown a quoit or paper aeroplane the furthest, and then measuring to check (use paces or footsteps), or seeing who has got nearest to the target in a game of boules or marbles (you could use cubes).

Don't pre-empt your child's thinking by pointing out that you and she have different-sized strides or feet. Let her work it out for herself.

Using rulers or tape measures

Involve your child in any measuring that you have to do around the house. Answer her questions as to why you are measuring, ask her to hold the end of the tape measure still for you and, as she gets older, to read off the numbers.

I want to know if this bookcase will fit in the alcove upstairs.

Weight

Playing

Your child will learn a lot about weight simply by playing with toys of different weights and discovering how they feel and behave.

Which block will make the balance go down, do you think? Let's try it and see.

A simple toy balance is valuable both for experimenting with measuring and for pretend play.

As your child gets older, you can make a game of putting a number of things in order according to weight, estimating first by feel, then weighing them on the balance to check.

Real weighing

Very young children can learn something from helping with real weighing, even if it is only the idea that weight can be measured. Let your child spoon the flour onto the scales; later he will be able to tell you when the pointer has reached a certain mark on the dial or, if you have balance scales, select the weights.

Volume and capacity

These usually go together in math learning: if a jug has a capacity of a quart, then the volume of liquid which will fit in it is a quart.

Water, sand and playdough

Playing with these is an excellent way to learn about volume and capacity. Just provide a selection of suitable containers, scoops for water and sand, and cutters and a rolling pin for playdough and let your child experiment.

Can you fit all the water in that jug into the bowl?
How many cups can you fill with the water in the teapot?

Don't interfere if children are enjoying themselves, but if they want some ideas, offer them a few challenges.

Area

At school children get their early ideas about area (the size of surfaces) from fitting shapes together to cover a surface and seeing how many are needed.

Tessellations

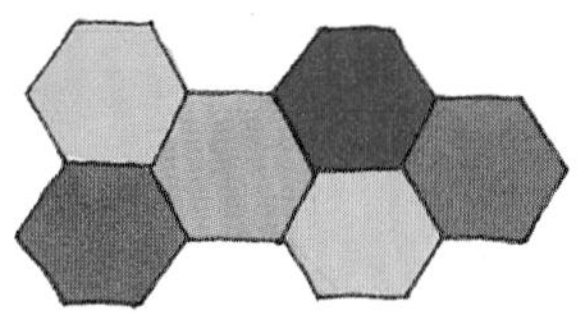

Hexagons tessellate.

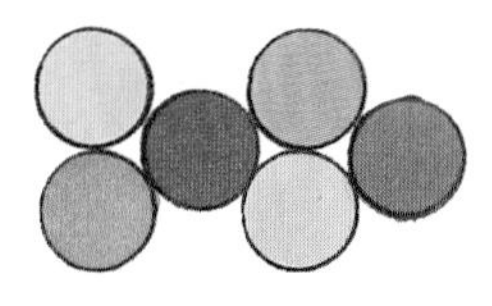

Circles do not.

Shapes which fit together without any gaps between them are said to "tessellate". Children sometimes enjoy creating tessellations. Try cutting out some shapes and seeing if they tessellate or not.

Once your child becomes aware of tessellations, she will probably spot them in all sorts of places, for example in fabrics, wallpapers, floor coverings, patchworks and brick walls.

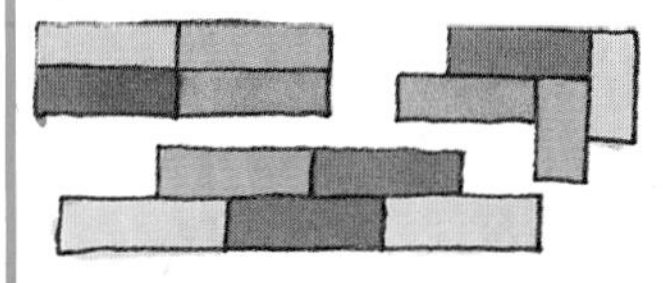

See if there is more than one way of tessellating a shape.

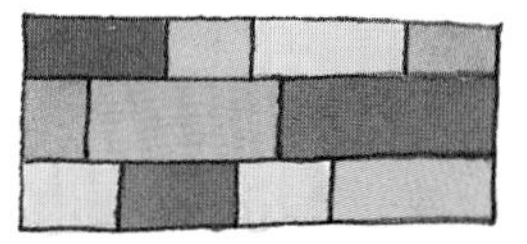

Can you find two shapes which tessellate together.

Packing

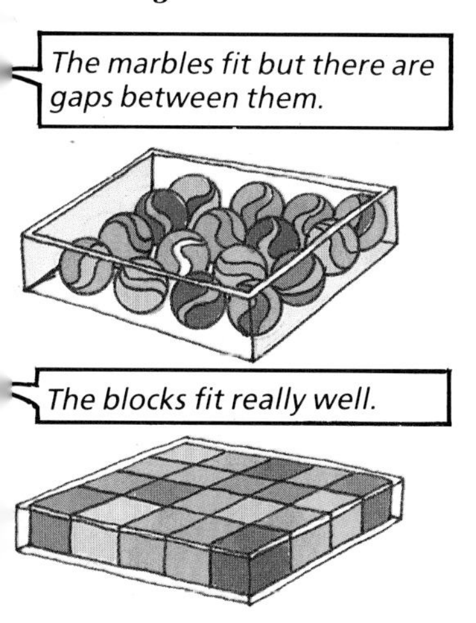

The marbles fit but there are gaps between them.
The blocks fit really well.

Packing things into containers and seeing how well they fit together will develop your child's ideas about volume and capacity.

Size and weight

This is lighter but it seems much bigger.

Children often get confused between size and weight because big things are so often heavier than light things. Get your child to shut his eyes while you put an object on each of his hands. Which is heavier and what are the objects?

Measuring volume and capacity

Is there enough milk for breakfast?

Estimating is particularly important when it comes to deciding how much of something there is or how much something will hold. Involve your child in this whenever you can as well as letting her pour liquids into the measuring jug.

You could print a birthday card or wrapping paper using tessellating shapes (see page 12 for materials to use).

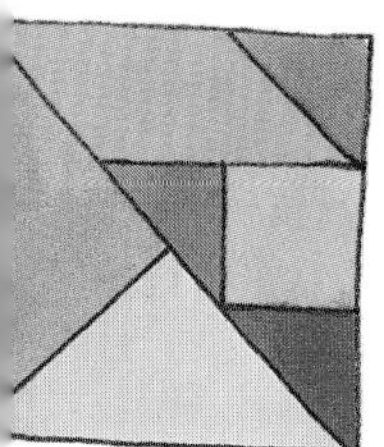

Jigsaws, mosaics and tangrams (Chinese puzzle consisting of a square cut into seven pieces) can also help to develop ideas about area. However you arrange the pieces, they have the same area.

Playing

Playing with dolls and dolls' houses gives plenty of opportunity for developing ideas about area.

Estimate whether there is enough material for a dress.

Fit blankets and tablecloths.

Paper walls.

Make carpets and rugs.

Measuring area

As your child gets older, involve her in real-life situations where you need to measure area, for example, to find out how many tiles you need, how big a carpet or how much material to make curtains or clothes.

TIME

There is more to time than telling the time by the clock.

Time is a difficult concept for children to grasp. They cannot see it or feel it, and the same period of time can seem very different from one day to the next depending on what they are doing.

Telling the time by a clock is a sophisticated skill which needs quite a lot of mathematical knowledge: the ability to count up to 60 in 5s, for example. Let your child take each stage very slowly and concentrate at first on helping him to develop a feel for the passage of time and to understand why we use clocks in the first place. This is as important as being able to tell the time accurately.

Talking about time

Days and weeks

We can't go to the park now because it's nearly lunch-time. We'll go this afternoon.

Many young children are vague about where they are in the day. Most are even vaguer about the days of the week. They realize that different days bring different events but are not sure of the sequence. Talking about "where you are" in time will help your child learn how time is divided up into morning, afternoon, evening, night; yesterday, today, tomorrow; and the days of the week.

Months and seasons

Is my birthday in summer?
No, it's in autumn, when the leaves start to fall off the trees.

The months and seasons are even harder for children to keep track of because they change so slowly. Try to relate them to specific events such as birthdays, holidays, or planting seeds in the garden and be prepared to answer the same questions repeatedly until your child gets things straight. Books showing sequences of events and seasons can help clarify ideas.

Making a record of time

Many children go through a phase of being extremely interested in time. Your child might like to make a days of the week chart, a calendar-cum-diary showing important events, a seasons picture (you can write the names of the months on it too) or, as he gets older, a chart showing what he does at different times of day. Incorporate drawings and photos in the charts.

Monday
Back to school
Tuesday
Swimming
Wednesday
Shopping
Thursday
Go to Sam's
Friday
'Wizzo' on Television
Saturday
Football
Sunday
Grandad's
1
Saw first snowdrops
2
My birthday
3
Library books due back
4
Granny came
5
Going to the zoo
6
Got chickenpox
7
Built snowman

Starting to use a clock

That hand doesn't move.
It does really. Let's see if it still points to the 4 after supper.

When your child starts showing an interest in clocks and watches, look at the different types with him. Talk about what the different parts do and point out that digital clocks and ordinary, analogue ones do the same job in a different way. Watch the minute hand going round on an analogue clock and the numbers changing on a digital one.

Lunch will be ready at 1 o'clock. That's when the long hand is pointing straight up, at the 12.

You can also start mentioning times very casually. With analogue clocks, it is best to start with the hours and then go on to the half and quarter hours, without worrying about the 5-minute intervals.* You can always talk about "nearly half past", "just after quarter to", and so on.

Hurry up, Mom! It's 11.20.
20 past 11 already! What time does the clock say?

It is easier for children to read off the time on a digital clock than to tell the time on an analogue one. They need to be able to use both sorts of clock though, so make sure, if your child has a digital watch, that you relate the time on that to the time on analogue clocks and vice versa.

How long is a minute?

To help your child get a feel for time and how it is measured, make a game of seeing what she can do in a minute while you watch the second-hand go round on a watch or count slowly to 60. Can she wash her hands or run round the garden? Let her time you too.

Do the same thing for a 5-minute interval using a kitchen timer or egg timer, or the minute hand on a clock. Can she tidy up her toys? Can you make a cup of tea?

Toy clocks

What time are we going?
At half past 2. That's when the hands on the real clock say the same as these.

Toy clocks with movable hands are useful, first for pretend play, later for learning to tell the time as well, provided you use them very casually.

Brass paper fastener
Cut-out cardboard hands

You may be able to give your child an old, real clock to play with. Or you could help her to make one from a paper plate, copying the face from a real clock.

Watching television and telling the time

Older children can be spurred to tell the time by looking to see when a favourite television programme starts, working out how long it lasts and checking it does not overlap with something someone else wants to watch. Children who are allowed to watch for a limited amount of time a day can work out what to watch without going over their limit.

* *For ideas on introducing fractions, see page 34. For counting in 5s, see pages 35, 36 and 37.*

MULTIPLYING AND DIVIDING

When children start doing multiplication and division at school, they do a lot of practical work involving adding up groups or "sets" of the same number (multiplying), and taking away groups or sets of the same number and sharing things out (dividing). They still learn multiplication tables but generally less by rote than children used to, more through their practical work, and games and puzzles. The greater understanding they have of the processes involved, the more likely they are to remember the tables and be able to use them sensibly.

3 rows 4 → 12
4 rows 4 → 16
9 bricks make
We'll need three potatoes each. How many is that altogether?
Are there enough sandwiches for us to have four each?
Can you share out the cherries fairly between the two of you?

Early work at school

This involves activities like those shown below. The children are encouraged to talk about what they have done, and perhaps to make a record of it in words, numbers and pictures.

Colouring rows of squares and working out the total.

Seeing how many sets of, say, 4, they can make from a pile of blocks.

Sharing out 10 pencils among the children in their group.

Children can learn a lot about multiplication and division by using calculators. For more about this see pages 41-42.

What to do at home

The best way to build up your child's skill at multiplying and dividing is to involve him in any multiplying and dividing you have to do yourself.

Fractions

Sharing things out, especially food, soon leads children to realize that numbers will not always divide exactly. You will find opportunities for cutting things up into halves, quarters and thirds and so introducing these words quite naturally.

Children can also grasp these simple fractions from folding, cutting or colouring in paper. Doing this with a circle of paper can make a useful link with learning to tell the time (half and quarter hours).

When children are older and have learnt to write fractions at school, they might like to see where these come on a number line.

Multiplication tables

Some schools still use lists of tables but many prefer to use a chart like the one below for teaching multiplication facts or bonds. Because each bond is shown twice, children can see that there are really not that many to be learned.*

To find 6 x 5, look for the square where the row starting with 6 meets the column headed 5. Or look for where the column headed 6 meets the row starting with 5.

The chart can also be used to do division. To find 28 divided by 4, for example, find the row starting with 4, look along it until you find 28 and the answer is at the top of the column.

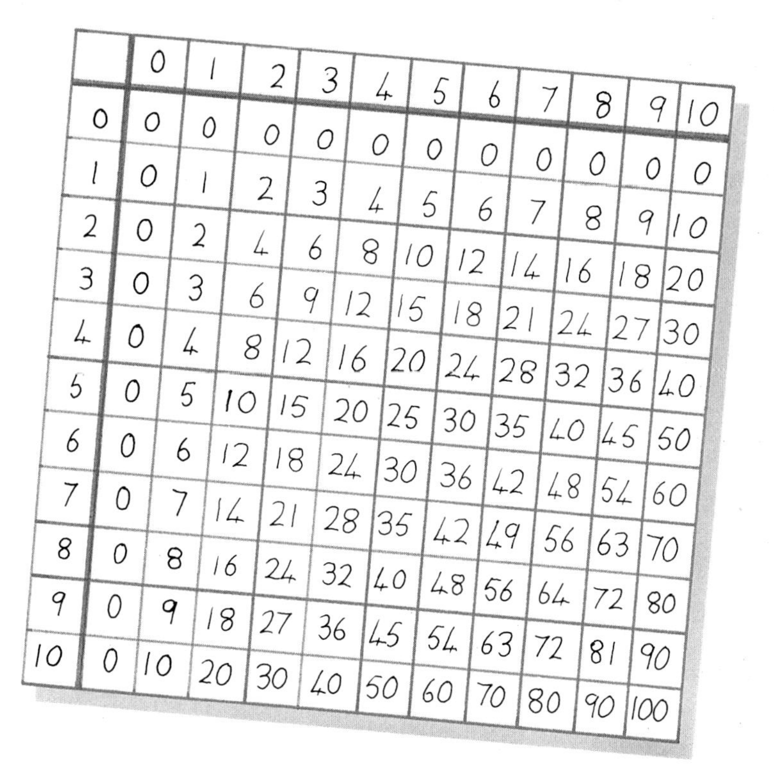

	0	1	2	3	4	5	6	7	8	9	10
0	0	0	0	0	0	0	0	0	0	0	0
1	0	1	2	3	4	5	6	7	8	9	10
2	0	2	4	6	8	10	12	14	16	18	20
3	0	3	6	9	12	15	18	21	24	27	30
4	0	4	8	12	16	20	24	28	32	36	40
5	0	5	10	15	20	25	30	35	40	45	50
6	0	6	12	18	24	30	36	42	48	54	60
7	0	7	14	21	28	35	42	49	56	63	70
8	0	8	16	24	32	40	48	56	64	72	80
9	0	9	18	27	36	45	54	63	72	81	90
10	0	10	20	30	40	50	60	70	80	90	100

You can use the chart as a checklist, in a similar way to the addition bonds chart on page 23, crossing off the bonds when you are sure your child knows them.

Other multiplication charts

Other types of chart may also be used at school for helping children learn multiplication bonds. Children can discover the pattern of a certain times table by colouring in all the numbers in that table on a chart, like the one shown below for the 3 times table.

1	2	3	4	5	6	7	8	9	10
1	12	13	14	15	16	17	18	19	20
1	22	23	24	25	26	27	28	29	30

Learning tables by heart

Unless your child positively enjoys chanting her tables parrot fashion and wants you to listen to her, it is best not to get her to do it. She may like you to test her at random, though, but make sure you let her test you as well, using the chart to help her if necessary.

* *Children in Britain used to learn tables up to the 11 and 12 times because of the British monetary system and the imperial system of measurement but this is no longer strictly necessary.*

Helping with tables

You can give your child practice at the 2 times or doubles table in various ways.

Show me 7 fingers. . . .
And here are 7 of mine.
How many is that altogether?

Use your fingers.

Share things out.

Give yourself 3 grapes, and give me 3. How many do we have altogether?

Six, eight.

Go up the stairs two at a time, counting as you go.

How many socks are there in the basket? So how many pairs must there be?

Count things that come in pairs.

5 times 7? Well, 10 times 7 is 70. Half that is 35.

The 5 times table is not difficult because of the way all the numbers end in 0 or 5, but if your child is not sure of it, he can check his answer by halving the 10 times bonds.

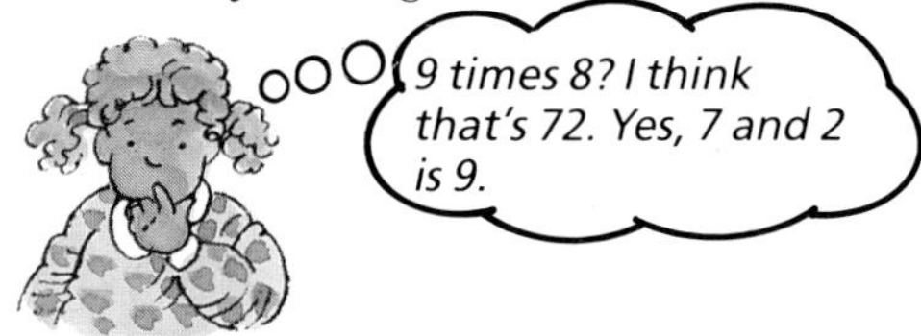

9 times 8? I think that's 72. Yes, 7 and 2 is 9.

She can check her answers to 9 times questions by adding up the two digits of the answer. These always add up to 9.

7 times 8? I know 7 times 4 is 28. Double that makes 56.
7 8s? I know 8 8s are 64. Take away 8 is 56.

The best way to help your child with multiplication bonds is probably to encourage her to be inventive about working out bonds she does not know.

Talk about the ways you each work out sums you do not know. Once she realizes she can work out bonds for herself, she will gain self-confidence and, with practice, will start to remember more and more of them. Always give her time to come up with an answer. If you rush her, she will panic and stop thinking properly.

Practical sums

As children get older, they can join in with working out multiplication and division problems around the home.

- Work out together how long a car journey should take you if you know the distance and the average speed you will be travelling.
- Using money is an excellent opportunity for putting multiplication bonds to use. If you need a certain number of things and they cost a certain amount each, how much will you spend altogether? How many things can you buy if one costs a certain amount and you have a fixed amount to spend?
- Let children help with working out how much food is needed for a certain number of people and then with sharing it out at mealtimes.
- Involve them in working out what is needed for do-it-yourself jobs, for example, how many rolls of wallpaper, how many packets of tiles?

Games to help teach multiplication bonds

Snap

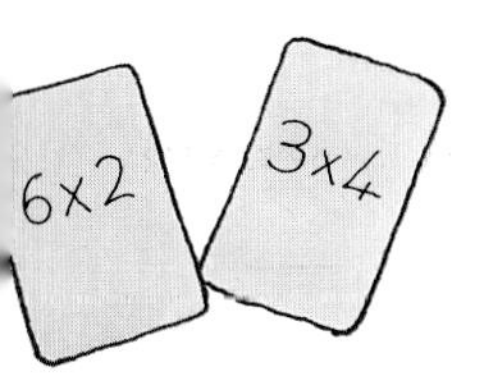

Play as for Snap II on page 25 but using cards showing multiplication or division bonds and their answers.

Pairs

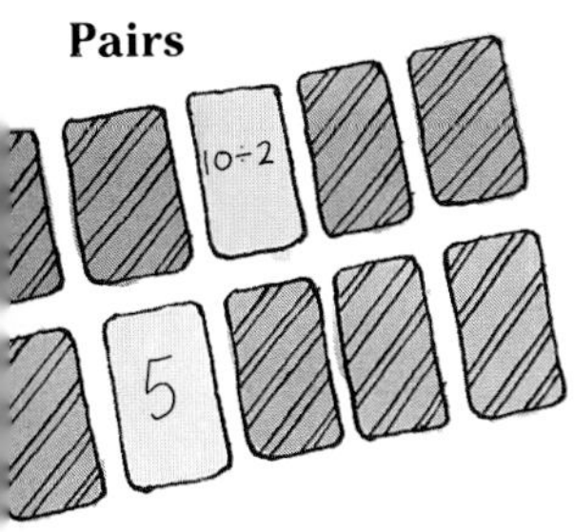

Play as for Pairs II on page 25 but using cards showing multiplication and division bonds and their answers.

Grids

Draw a grid as shown below. Take it in turns to throw a dice and cover with a counter a multiple of the number thrown. For example, if you throw a 3, you can cover the 3, 6, 9, 12, 15, 18, 21 or 24. Let children use a calculator to help them if necessary. The first to make a row of three counters is the winner.

Variation

- Make a grid showing bigger numbers. Throw a pair of dice, add up the two numbers and cover up multiples of the total; or cover a dice with labels showing the numbers of any tables your child especially needs to practise.

Spiders

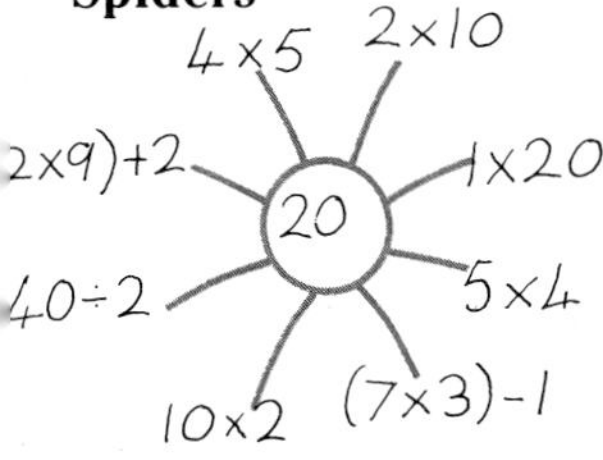

Take it in turns to give bonds which make up a certain number, as shown here. Include addition and subtraction too.

Number line jumps

Number lines are used in schools for multiplying and dividing as well as adding and subtracting.

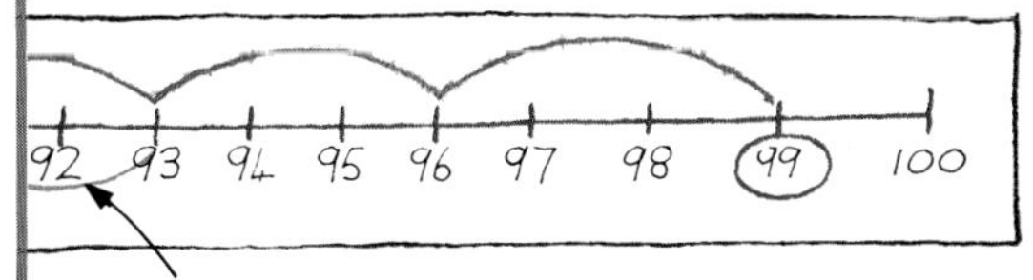

One person should draw above the line and one below it.

Choose a table that your child needs to practise, for example the 3 times. Take turns to throw a dice and draw that many jumps of 3 along the line, starting from 0. In this case, you will finish on 99 if you are playing on a line to 100. The first to reach the end exactly wins.

Variation

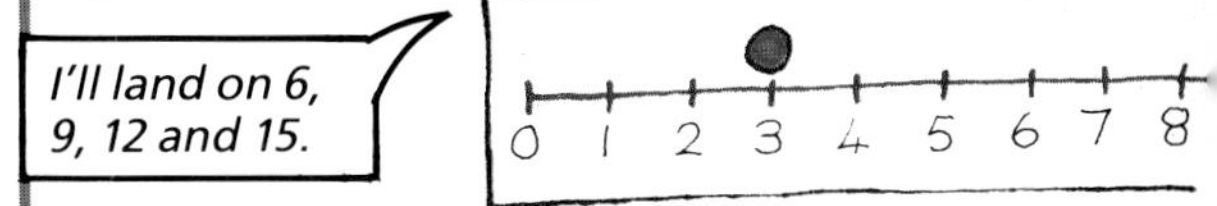

- Move a counter along the line instead of drawing your jumps. Encourage your child to remember the landing points.

Buzz

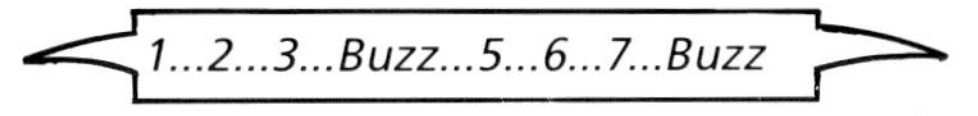

Decide on a table that you want to practise. Then start counting, taking it in turns to say each number. When you get to any number in the table you have chosen, say "buzz" instead of the number.

Variations

34...Buzz...32...31...Buzz

- Count backwards.

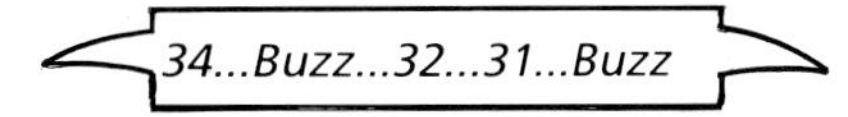

- Count in 2s.

1...2...Fizz...4...Buzz...Fizz...7

- Play Fizz-buzz. Numbers in the 3 times table are a "fizz" and in the 5 times table a "buzz".

MENTAL ARITHMETIC

Feeling confident about mental arithmetic is extremely important. It is often quicker and easier to do a mental calculation than to fetch a pencil and paper, or a calculator. Even if children do use one of these aids, they still have to be able to do some of the working in their heads. First, they have to decide what sort of sum they need to do. When they write a sum down, they are really just using the paper to keep a record of what they are thinking through in their heads. When they use a calculator, they have to be able to judge whether the answer the calculator gives is a sensible one and find some way of checking it.

The games and activities suggested throughout this book, as well as those opposite, will give your child useful practice in mental arithmetic. One of the best ways of all to help your child gain confidence at manipulating numbers in his head is to talk with him about the methods you both use to do mental calculations.

Comparing methods*

Your child may be encouraged to try out new ways of doing mental calculations if you compare the methods you both use. Always make it clear that there is more than one "proper" method: it is better for her to use a slow, relatively inefficient method that she understands than a better method that she does not. Be prepared to admit that your own methods too could perhaps be improved.

Don't pressurize your child to come up with answers to sums quickly. Many adults with unhappy memories of quick-fire mental arithmetic sessions at school still experience feelings of panic when faced with a mental calculation; this actually prevents them from doing the sum as quickly as they could.

How do you work out 17 plus 9 plus 3?... I add 17 and 3 to make 20, and another 9 makes 29.
17 add 10 makes 27, take off 1 makes 26, add on 3 makes 29.
How do you subtract 7 from 15?... I take off 5 which is 10, then take off another 2 which makes 8.
The difference between 7 and 10 is 3; add on 5 makes 8.
How do you work out 13 times 4?... I double 13 to make 26, then double 26 to make 52.
I know that 4 times 10 is 40, and 4 times 3 is 12; 40 add 12 is 52.
How do you work out 29 divided by 3? I think that 8 3s are 24, so 9 3s are 27 and there's 2 left over.
I know that 10 3s are 30 but that's too much; so 9 3s must be 27, and there are 2 over.

* *See page 36 for comparing methods to help learn multiplication bonds.*

Calculators and mental arithmetic

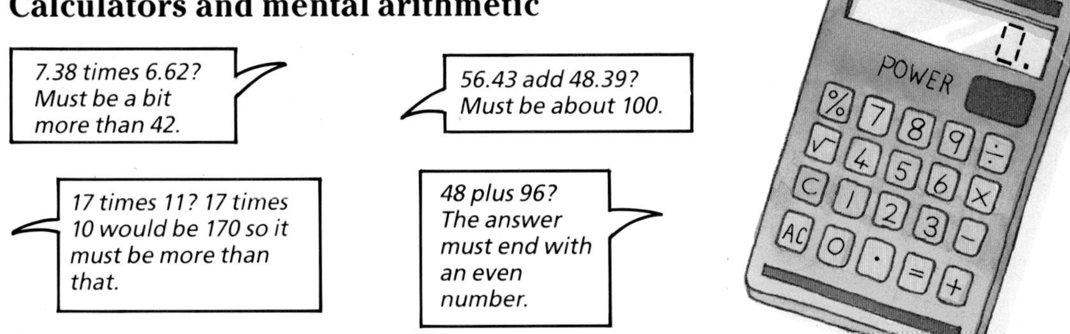

7.38 times 6.62? Must be a bit more than 42.
56.43 add 48.39? Must be about 100.
17 times 11? 17 times 10 would be 170 so it must be more than that.
48 plus 96? The answer must end with an even number.
POWER
0.

It is very easy to press the wrong button on a calculator by mistake so it is essential to have some way of checking whether the calculator's answer is likely to be correct. This means doing a rough estimate in your head.

Games

Guess and count*

Learning to make sensible estimates is as important as doing accurate calculations. Ask your child to guess how many nails there are in a packet or pieces of pasta in a jar and then help him count them to see how close his estimate was.

Help him to make sensible guesses by looking at a group of 10 or 100, then working out roughly how many of these groups there must be.

If he gets lost with his counting, help him to organize the objects into 10s or 100s and then to count these.

Car numbers

Encourage your child to play with numbers in her head. For instance, if you are stuck in a traffic jam, try looking at a nearby number plate and trying to connect the numbers in some way.

16 is 4 4s. So 160 is 4 40s.

160 JMB

Totals

You need one suit of cards per player. Shuffle the cards and deal them out. Each player uses their cards to make as many sets totalling 20 as they can. (An ace is worth 1; all the court cards are worth 10.) The person with the fewest cards left over at the end wins.

Variation

- Aim for another total such as 15, 21 or 27.

Cards left over

* *This game is also good for increasing understanding of place value (see page 26).*

USING A CALCULATOR

Many parents worry that if children are taught to use calculators, they will be unable to do any sums without one.

In fact, calculators can develop children's mathematical thinking rather than limit it. Besides enabling them to do complex calculations quickly, calculators allow children to concentrate on the mathematical ideas behind problems without getting bogged down in arithmetic which may be so difficult as to prevent them from coming up with an answer at all without a machine. Calculators are also invaluable for allowing children to explore and play around with numbers.

Children have to be quite knowledgeable in the first place to use a calculator for doing sums. They have to know what type of sum to do, which keys to press and, very importantly, whether the answer the calculator shows is likely to be correct. (There is more about this on page 39.)

On these pages are just a few ideas for things to do with a calculator. They will help to develop your child's all-round number skills besides building confidence in using a calculator. For more ideas, look for books of calculator games and activities in the shops.

Calculators and young children

Let children play with a calculator when they are still quite young. They can learn to recognize the numbers on the display (these will look different to them from numbers in books). They can learn which keys clear the screen, which keys make letters appear and can become familiar with the symbols +, −, ×, ÷ and = even though they don't yet know what they mean. All this will give them confidence when they come to use calculators "properly" later.

What's that number?
Five million, sixty thousand and seventeen.
That's a zillion.
Is it? You know, some numbers are so big they can't fit on the display. I'll write you one that won't fit.

Answer your child's questions about the calculator but don't tell her she is wrong too often or she will get the idea that calculators are too difficult for her.

Making numbers appear

For young children it can be quite a challenge just to get the display to show what they want it to.

Write down a number and see if your child can enter it correctly.

Can she enter numbers when you say them in full?

Adding 1

Keep adding 1 on the calculator. Can your child predict what numbers will appear in each position each time? Can she say the correct name for the whole number?

Variation

- Try adding 2s or 5s or 3s.

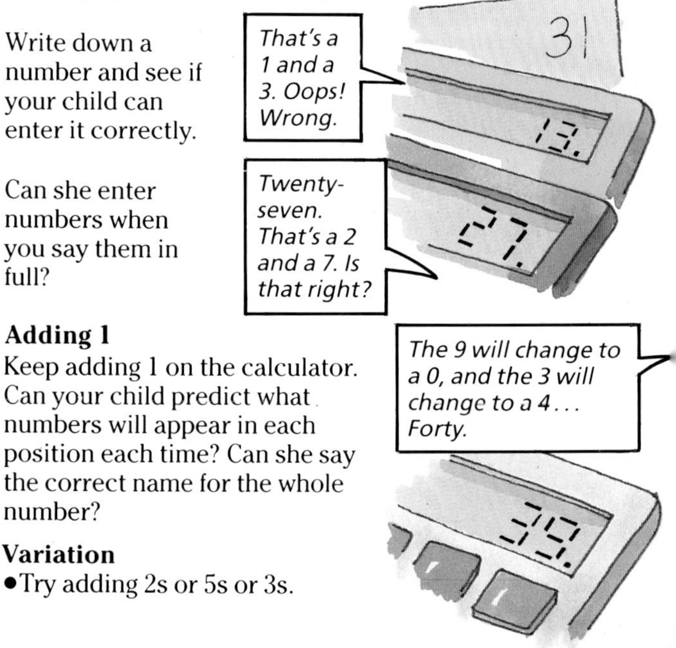

31
That's a 1 and a 3. Oops! Wrong.
13.
Twenty-seven. That's a 2 and a 7. Is that right?
27.
The 9 will change to a 0, and the 3 will change to a 4... Forty.
39.

Activities for older children

Making numbers appear

Can your child make a number appear without pressing the digits in that number? For example, for 32, he might do the following:

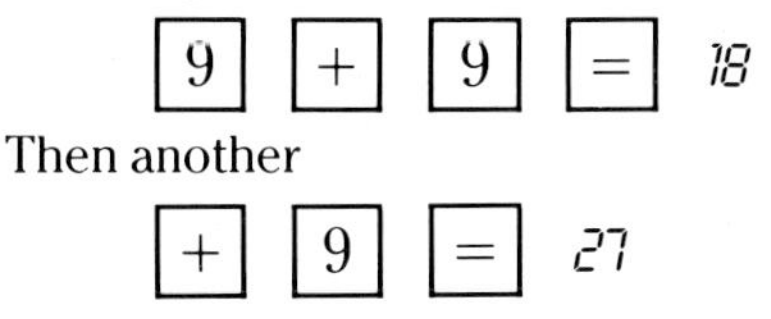

And another

+ 9 = 36

That's too much. And he's not allowed to do −2 or −3. So he tries:

− 4 = 32

Own goal game

The goal is 29 and you don't want to be the person to reach it. Set the calculator at 0 and take it in turns to add any number from 1 to 9 inclusive. The person who reaches 29 is the loser.

Variations

- Choose a different goal.
- Toss two dice and choose which of the numbers thrown you want to add.

Making numbers disappear

Can your child make numbers disappear without pressing the digits in that number? So, for 56 for instance, he can't press 5 or 6. So he tries:

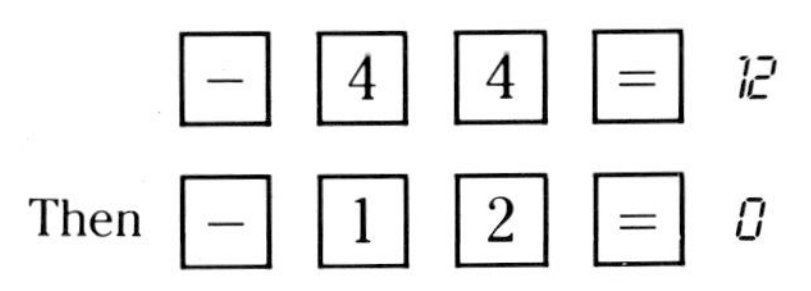

If he is allowed to use the digits in the number, can he make the number disappear by removing each part of it (the 10s and units) separately and in one go? So, for 56:

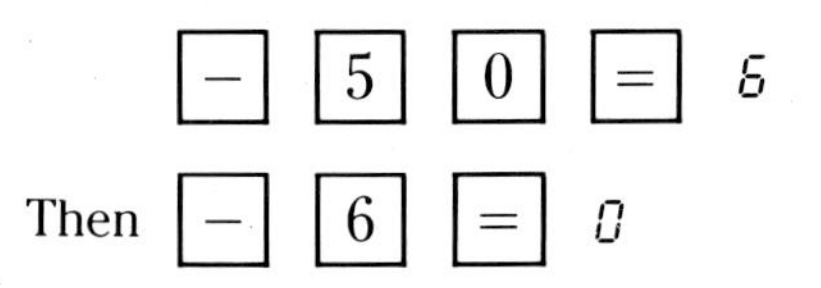

Adding 1s, 10s and 100s

When your child is quite confident about predicting what number will come up next on the display when you add 1, try adding 10s or 100s repeatedly. Can she still predict the changes?

This activity will get your child looking at the different digits in a number and thinking about their place value (see page 26). It should help her realize that if you add a certain number of 1s to a number, it will only change a little but that if you add that many 10s to a number it changes a lot.

Variation

- Start at 100 and subtract 1s, or 10s.

1s and 10s game

One of you chooses a target number below 50, the other a target number between 50 and 100. Put the calculator on 50.

Take it in turns to throw a dice and then decide whether to add or subtract that many 10s or that many 1s. The aim is to get as close to your target as possible. (If you make the calculator go below 0 or over 100, you lose.)

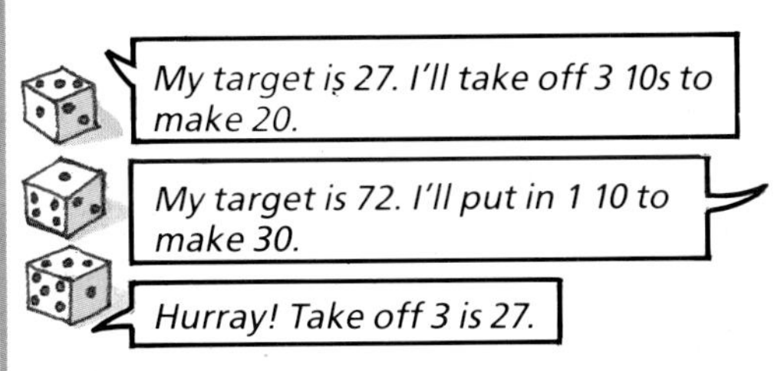

My target is 27. I'll take off 3 10s to make 20.
My target is 72. I'll put in 1 10 to make 30.
Hurray! Take off 3 is 27.

Multiplication

Calculators are useful for reinforcing work done on multiplication bonds. Keep adding a certain number. Can your child predict what numbers he will come to?

There aren't any 5s or 7s.
0 2 4 6 8
10 12 14 16
18 20

Suggest he writes down the numbers. Can he see a pattern? What digits do they end with? What don't they end with?

Decimals

Children often meet decimals for the first time through playing with a calculator. To understand them your child will need to have a good grasp of place value (see page 26).

The idea can be reinforced by watching the numbers clocking up on a car distance meter (the right-hand number usually shows tenths of a kilometre or mile); by doing sums with money; or by making a number line showing decimals.

Decimal tricksy

45·3 is longer than 46 but is it bigger?
45·3
46
5·2

Try playing a game of Tricksy (see page 19) but include in your deck some cards showing simple decimal numbers.

Negative numbers

Negative numbers will sometimes appear when a child is playing with a calculator. Your child may already have begun to think about these, perhaps from looking at a thermometer and talking about temperatures "below zero".

Try helping her to extend a number line to include negative numbers.

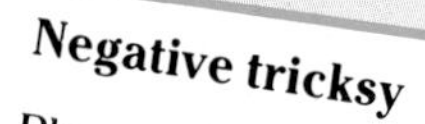

Negative tricksy

Play a game of Tricksy (see page 19) including in your deck some cards showing negative numbers.

Calculator jumps game

Play as for Number line jumps on page 37 but use a calculator each instead of a number line.

Three hops to 100

Choose any number under 100 and try to get from there to 100 in three "hops" or fewer. For each hop you must press:

one of the keys +, −, ×, ÷,
one number key (0-9 inclusive),
and the = key.

For 37, for example, you could do:

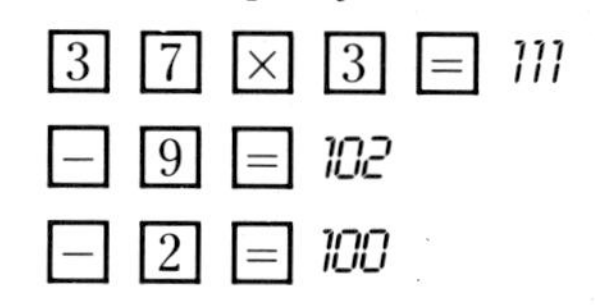

4s only

What numbers can you make the calculator display using only the keys 4, +, −, ×, ÷ and =? Can you make all the numbers up to 10? For example, to get 6, you could do:

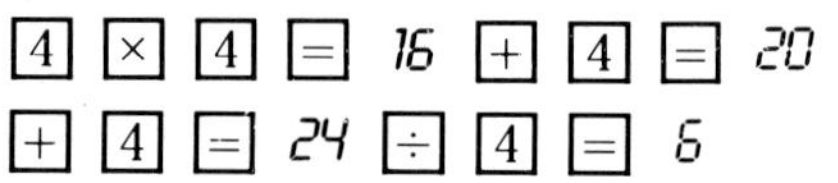

What's your date of birth?

Here is a trick for working out when someone was born. Without your seeing, ask them to enter into a calculator the day of the month on which they were born. Say their date of birth is 2 October 1982. They enter 2. Then tell them to do the following:*
Multiply by 20, add 3 and multiply by 5.

[2] [×] [2] [0] [+] [3] [×] [5]

Add on the number of the month they were born in and again multiply by 20, add 3 and multiply by 5.

[+] [1] [0] [×] [2] [0] [+] [3] [×] [5]

Then add on the last two figures of the year in which they were born.

[+] [8] [2] [=] 22597

To work out their date of birth, take the calculator and subtract 1515 from the number in the display. Then if you read the figures in the display from left to right you have the day, month and year of birth.

[−] [1] [5] [1] [5] [=] 21082

* *If you have a scientific calculator you may have to press the = key after each part of this calculation.*

COMPUTERS AND MATH

It is worth letting children use computers from an early age, not only because they will need to use them at work and probably at home when they grow up, but because they can be of enormous educational value, helping them to understand quite difficult ideas at a much younger age than previously.

For example, children can now do far more sophisticated work on graphs at school than they used to. They can get the computer to do the drawing for them and this leaves them time to concentrate on examining the different types of graph and how they work, and on deciding which is most suitable for their needs.

Most primary schools have computers but the way they use them will vary depending on resources. If you have a computer at home, your child can certainly benefit from using it, though you will need to encourage her to use it productively and not just for playing arcade, space-invader-type games.

Logo

This is a programming language which has been specially designed for children. It is mainly used for producing graphics on the screen and is probably the single richest activity available to children on the computer.

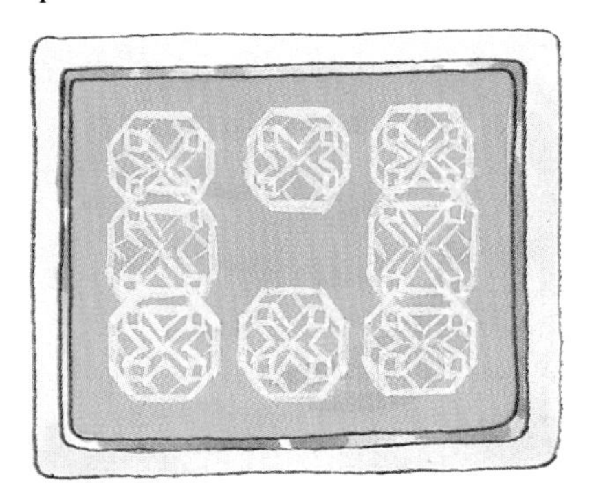

Even young children can use and enjoy Logo, making simple shapes or pictures; older children can use it in increasingly sophisticated ways, making spirals and complicated moving patterns, and even to do calculations.

The way you get the computer to draw on the screen is by instructing a small marker, called a "turtle", to move in a straight line and turn through angles whatever distance you choose. This means that children get experience of estimating and calculating angles and distances, think about shapes and patterns and learn to handle large numbers with confidence.

There is a form of Logo available for most makes of home computer. One way of using it together with your child is by taking it in turns to give each other instructions for the turtle.

Math adventure games

Children can be motivated to use computers by playing adventure games. The puzzles or problems that players have to solve during the course of these games are often of a mathematical nature. Your child's school might be able to suggest games suitable for home use. Otherwise, look in a computer shop or magazine.

Games of strategy

Children often find games of strategy such as Chess, and Noughts and Crosses very interesting. There are computer programs for these and because the games are played alone, children have time to concentrate fully on the strategies involved, without feeling hurried to make a move by another player.

THE TEACHING OF MATH

The way that math is taught in schools has changed considerably over the years, as the needs of society have changed and more has been discovered about how children learn.

Calculators and computers

Before the days of calculators and computers, large numbers of people were needed who could perform routine calculations accurately and, in certain jobs such as book-keeping, neatly in "longhand".

Nowadays, greater emphasis is placed on helping children to understand underlying mathematical concepts rather than on practising mechanical skills like long division. Children still need to know how to calculate but they also need to understand the calculation processes if they are to be able to use math to solve problems in real life.

A natural desire to learn

It is obvious just from watching a toddler playing that children have a natural desire to learn. The old "chalk and talk" method of teaching, in which children remained fairly passive, left many of them feeling confused and alienated. It is now realized that children learn better by "doing" and much early math teaching revolves around giving children opportunities to explore and experiment in the classroom through play.

Don't make the mistake of thinking that your child is "only" playing or, because she is not doing what you did at school, that she is learning nothing. The teacher will channel the children's explorations by talking with them about what they have found out and helping them to ask questions which will lead to further discoveries.

As they get older, they will start to do more formal work as well. Much of this will still be practical but they will probably start to make some record of their work in words or pictures.

Listening to children

Watching what a child does and listening carefully to what she says helps adults to find out not only what she is interested in, but how she is thinking and what she does and does not understand. Teachers often use the tactic of watching and listening to find out what children need to learn.

For example, suppose a group of children are playing a board game. One of them is on 10 and needs to move on 8 steps but instead of just thinking "That's 18" she laboriously counts on 8. The teacher might ask the children if any of them can find a quicker way. If they cannot, she will make a note to plan some work for them where adding numbers to 10 is the main focus.

At home, it is important to adopt an even more "softly, softly" approach. If you provide interesting and stimulating materials, your child will learn an enormous amount simply through playing.

Playing with children

It is vital to allow children time to play with toys on their own and in their own way. When you are playing with your child, try not to be too directive. Let her take the lead and follow her ideas as far as is reasonable. The time to make your own, casual suggestions is when she is familiar with the materials.

Resist all temptation to take over, for example, pointing out a jigsaw piece when she is still quite happy looking for it herself. She will learn better and develop greater self-confidence the more she is able to do for herself. When she asks for help, try to give guidance and clues rather than providing the answer straightaway.

Don't be surprised if your child doesn't learn what you expected her to from a particular activity. Teachers are very used to the phenomenon of young children missing the main point but being fascinated by some other, equally worthwhile aspect.

Talking with children

The way adults talk with children is important too. If an adult is interested in what a child does and wants to talk with her, the child will gain self-confidence.

The way you use words with your child will help her to learn new ways of expressing herself and can increase her understanding of basic mathematical concepts. Make a point of using words to describe attributes such as size and quantity even with a very young child.

Math must be relevant

Children learn best when a subject has real meaning for them. In math this means doing a lot of practical work, where they learn skills "on the job". For instance, in school, children might help to plan an end-of-term party, working out quantities of food and costings, and planning how to organize the time available. With young children, it means counting the number of raisins or grapes they are allowed to eat rather than counting things in a book as an educational exercise.

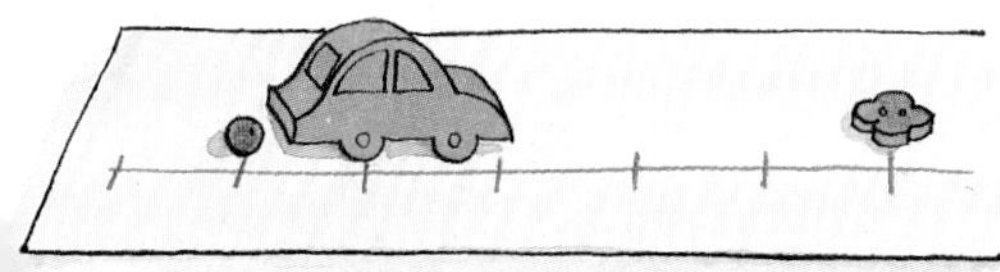

Enjoyment is important

Children learn if they enjoy what they are doing. This is why teachers use so many games in school both for introducing new ideas and for practising and developing skills. For instance, if a teacher wants children to think about what happens when 10 is added to a number, she might adapt Snakes and Ladders so that all the snakes take the players back 10 spaces and all the ladders take them forward 10. She might also provide two dice numbered only with 4s and 5s, so that children will quite often throw a 10. She will then ask them to predict where they will land before they move.

Math at your child's school

Most schools welcome parents' interest in what they are doing and some even have open evenings to explain their aims in particular subjects.

The rate at which individual children progress varies widely in math, as it does in other aspects of development. If at any stage you think your child is having difficulties with math, ask to discuss the problem with her teacher. Ask if there are any ways in which you could help at home, without pushing your child. Are there any activities your child has particularly enjoyed which you could follow up?

It is important always to try to work with the school as far as possible. If you are unhappy with the school's approach to math, discuss it with them but try not to let your child become aware of your dissatisfaction. It is difficult for young children to cope with differences of opinion between home and school.

Girls and math

It is thought that girls and boys start out with roughly equal mathematical ability. However, math has been traditionally thought of and presented as a boys' subject and far fewer girls than boys continue to study it beyond the stage where it is compulsory.

Many girls need much more encouragement with their math if they are not to miss out on all the opportunities that learning math provides. Parents and teachers need to ensure from the start that girls have just as much opportunity to play with construction toys, calculators and computers as boys. Boys tend to demand more of the teacher's time and attention in the classroom than girls do, so as your daughter gets older, you may find you need to play math-type games at times when her brothers are not around to dominate the proceedings.

...MBER SONGS AND RHYMES

Children can learn numbers effortlessly through songs and rhymes.

The songs and rhymes below and opposite are all useful for teaching numbers to young children. Remember, though, to use them primarily simply for enjoyment. Put actions to them when you can, holding up the appropriate number of fingers or counting on your fingers, for instance.

Some of the songs which start at "one" and work up through the numbers you may like to take only as far as "five" to start with. As your child gets older you can take them up to "ten". Similarly, songs that work backwards from "ten" can be started from "five" for very young children.

One, two, three, four, five,
Once I caught a fish alive.
Six, seven, eight, nine, ten,
Then I let him go again.
Why did you let him go?
Because he bit my finger so.
Which finger did he bite?
This little finger on the right.

One, two, buckle my shoe,
Three, four, knock at the door,
Five, six, pick up sticks,
Seven, eight, shut the gate,
Nine, ten, a good fat hen.

One potato, two potato, three potato, four,
Five potato, six potato, seven potato, more.

Peter hammers with one hammer, one hammer, one hammer;
Peter hammers with one hammer, this fine day.
Peter hammers with two hammers, two hammers, two hammers;
Peter hammers with two hammers, this fine day.
Continue up to "four hammers". Then:
Peter's getting tired now, tired now, tired now;
Peter's getting tired now, this fine day.
Peter's going to sleep now, sleep now, sleep now;
Peter's going to sleep now, this fine day.
Peter's waking up now, up now, up now;
Peter's waking up now, this fine day.

This old man, he played one,
He played knick-knack on my drum.

Chorus: insert at end of each verse.
With a knick-knack, paddy-wack, give a dog a bone,
This old man came rolling home.

This old man, he played two,
He played knick-knack on my shoe.
This old man, he played three,
He played knick-knack on my knee.
This old man, he played four,
He played knick-knack on my door.
This old man, he played five,
He played knick-knack on my hive.
This old man, he played six,
He played knick-knack on my sticks.
This old man, he played seven,
He played knick-knack up to heaven.
This old man, he played eight,
He played knick-knack on my gate.
This old man, he played nine,
He played knick-knack on my spine.
This old man, he played ten,
He played knick-knack on my hen.

One, two, three, four,
Mary at the cottage door,
Five, six, seven, eight,
Eating cherries off a plate.

Five currant buns in the baker's shop,
Round and fat with sugar on the top.
A little girl came to the shop one day,
She bought one bun and took it right away.
Four currant buns in the baker's shop...
Continue.

One little rabbit, wondering what to do,
One more came along and then there were two.
Two little rabbits, sitting down to tea,
One more came along, then there were three.
Three little rabbits, knocking at the door,
One more came along, then there were four.
Four little rabbits, going for a drive,
One more came along, then there were five.
Five little rabbits, getting up to tricks,
One more came along, then there were six.

Five little ducks went swimming one day,
Over the pond and far away.
Mummy duck said, "Quack, quack! Quack, quack!"
But only four little ducks came back.
Four little ducks went swimming one day...
Continue down to:
But no little ducks came swimming back.
No little ducks went swimming one day,
Over the pond and far away.
Mummy duck said, "Quack, quack! Quack, quack!"
And five little ducks came swimming back.

Five friendly frogs, sitting on a well;
One leaned over and down he fell.
Frogs jump high, frogs jump low,
Four friendly frogs jump to and fro.
Four friendly frogs sitting on a well...
Continue.

Five little speckled frogs,
Sat on a speckled log,
Eating some most delicious bugs,
Yum, yum.
One jumped into the pool,
Where it was nice and cool,
Now there are four more speckled frogs,
Glug, glug.
Four little speckled frogs...
Continue.

Five little monkeys jumping on the bed,
One fell off and bumped his head.
Mummy phoned the doctor and the doctor said,
"No more monkey business, jumping on the bed."
Four little monkeys jumping on the bed...
Continue.

Five fat sausages, frying in a pan,
All of a sudden, one went "Bang!"
Four fat sausages, frying in a pan...
Continue.

Ten pink parrots, sitting on a wall,
Ten pink parrots, sitting on a wall,
And if one pink parrot should accidentally fall,
There'd be nine pink parrots, sitting on a wall.
Nine pink parrots...
Continue.

There were ten in the bed,
And the little one said,
"Roll over, roll over."
So they all rolled over,
And one fell out,
There were nine in the bed,
And the little one said,
"Roll over, roll over."...
Continue down to:
There was one in the bed,
And this little one said,
"Thank goodness, peace at last."

One man went to mow, went to mow a meadow,
One man and his dog, went to mow a meadow.
Two men went to mow, went to mow a meadow,
Two men, one man and his dog, went to mow a meadow.
Three men went to mow...
Continue.

INDEX

First published in 1989 by Usborne Publishing Ltd, 20 Garrick Street, London WC2E 9BJ, England.